PLACE IN RETURN BOX to remove this checkout from your record.
TO AVOID FINES return on or before date due.
MAY BE RECALLED with earlier due date if requested.

DATE DUE	DATE DUE	DATE DUE

5/08 K:/Proj/Acc&Pres/CIRC/DateDue.indd

The Riverine Ecosystem Synthesis
Toward Conceptual Cohesiveness in River Science

AQUATIC ECOLOGY Series

Series Editor

James H. Thorp
Kansas Biological Survey
University of Kansas
Lawrence, Kansas

The Riverine Ecosystem Synthesis

Toward Conceptual Cohesiveness in River Science

James H. Thorp

Senior Scientist
Kansas Biological Survey
University of Kansas, Lawrence, Kansas USA
Professor
Department of Ecology and Evolutionary Biology

Martin C. Thoms

Professor
Riverine Landscape Research Laboratory
University of Canberra, Canberra, ACT, Australia

and

Michael D. Delong

Professor
Large River Studies Center and Department of Biology,
Winona State University, Winona Minnesota USA

Amsterdam • Boston • Heidelberg • London • New York • Oxford
Paris • San Diego • San Francisco • Singapore • Sydney • Tokyo

Academic Press is an imprint of Elsevier

Academic Press is an imprint of Elsevier
84 Theobald's Road, London WC1X 8RR, UK
Radarweg 29, PO Box 211, 1000 AE Amsterdam, The Netherlands
Linacre House, Jordan Hill, Oxford OX2 8DP, UK
30 Corporate Drive, Suite 400 Burlington, MA 01803, USA
525 B Street, Suite 1900, San Diego, CA 92101-4495, USA

First edition 2008

Notice
No responsibility is assumed by the publisher for any injury and/or damage to persons
or property as a matter of products liability, negligence or otherwise, or from any use
or operation of any methods, products, instructions or ideas contained in the material
herein. Because of rapid advances in the medical sciences, in particular, independent
verification of diagnoses and drug dosages should be made

ISBN: 978-0-12-370612-6

For information on all Elsevier publications
visit our website at books.elsevier.com

Printed and bound in USA
08 09 10 11 12 10 9 8 7 6 5 4 3 2 1

Working together to grow
libraries in developing countries

www.elsevier.com | www.bookaid.org | www.sabre.org

ELSEVIER   BOOK AID
            International   Sabre Foundation

Contents

4   *The Spatial Arrangement of River Systems: The Emergence of Hydrogeomorphic Patches*

5   *Defining the Hydrogeomorphic Character of a Riverine Ecosystem*

6   *Ecological Implications of the Riverine Ecosystem Synthesis: Some Proposed Biocomplexity Tenets (Hypotheses)*

7 Ecogeomorphology of Altered Riverine Landscapes: Implications for Biocomplexity Tenets

8 Practical Applications of the Riverine Ecosystem Synthesis in Management and Conservation Settings

Foreword

Upon encountering documents with ambitious to audacious titles that are pointed about things we take very seriously, we scientists have a natural and understandable tendency to first look at the literature cited to see if our papers have been included. This may be especially true for a book that claims to be a synthesis of our science, in this case river ecology. When you read this book initially, don't do that. Try to leave your personal interests and pet theories aside for your first run through this essay; it is intended to make you think synthetically – and it will. Let the information and ideas flow like water through the interactive and hierarchical habitat patches that compose the river and its flood plain, checking for retention and transformation processes as you go.

Yes, this book is about all, or at least most, of the things we hold near and dear in the ecology of running waters. And you will need to read it a couple of times and keep it close for reference, because there is a great deal of information and some profound ideas in it. For those that are well-read in river ecology, the 17 or so interactive, central tenets will not surprise you much initially, but you will have to agree that they are unifying of ideas we have discussed independently for years. For those that are not well read – but are serious about river ecology – this book is essential reading and will be thought-provoking.

A very important feature of the book is that it is a novel convergence of ideas that emerged from river ecological studies across continents and, especially, across latitudes. Like most Americans and Europeans with our north temperate, and all too often small stream biases, I have entertained the thought that Australian and tropical rivers were too different or perhaps too poorly understood to precisely fit a general view of riverine structure and function. Herein the authors explode that view with a logical analysis of theory and practice that in fact is a riverine ecosystem synthesis that does apply generally and also specifically to your favorite river.

The book is not a complete synthesis, in part intentionally in that they leave out the vertical dimension of river ecosystems for lack of, they say, expertise on surface and ground-water interactions. Moreover, the book must necessarily be a "starter" for development of novel new hypotheses because rivers and their catchment basins encompass the enormously complex biogeochemistry of the continents, with the confounding of human influences stacked pervasively and haphazardly on top. The Riverine Ecosystem Synthesis is a great contribution in helping us think holistically about rivers and their biota from organismal to landscape levels of organization. It remains for us to use the lessons and implications proactively to enhance human well-being.

Jack A. Stanford
Flathead Lake Biological Station
The University of Montana

Preface

The impetus to write a book on the riverine ecosystem synthesis emerged at the 2005 annual meeting of the North American Benthological Society in New Orleans, and barely 2 months later, we signed a contract with Academic Press. This book was to be an expansion of a manuscript that was *In Press* at that time in *River Research and Applications* (Thorp *et al.*, 2006). However, the true origin of this synthesis, journal publication, and book was a rivers meeting held in Albury, NSW, Australia, in July 2003 where Jim gave a plenary talk (with suggestions from Mike) at the request of Martin, the conference leader. Martin had asked Jim to speculate and not to worry about being controversial – he got his wish! Shortly after the meeting, the three of us joined together to write a conceptual paper that greatly expanded the hypotheses presented at that meeting.

Three important goals of our symposium talk, journal article, and book have been to (i) develop some measure of conceptual cohesiveness for the study of riverine landscapes by synthesizing crucial elements of the many lotic ecology models published from 1980 to the present along with those of landscape ecology and fluvial geomorphology; (ii) present a new perspective on how riverine landscapes are physically and ecologically structured along longitudinal and lateral dimensions; and (iii) integrate approaches from small to large spatiotemporal scales throughout the riverine landscape as a framework for research. A fourth goal emerged during discussions of the book itself – making theory for riverine landscapes both easy to apply by practicing ecologists/environmental scientists and useful for studying the significantly altered rivers found in most countries. This last goal has expanded to include recommendations for river management, monitoring, and rehabilitation.

Initially a name for this conceptual approach was avoided because we wanted to emphasize that it was a synthesis of many theories rather than strictly a new model. Although we supported the idea of conceptual and mathematical modeling in aquatic ecology, we agreed with some critics that there were too many small scale models, theories, and purported paradigms in the scientific literature, each with a different name but few integrated with other models. Indeed, the initial *In Press* copy of our journal manuscript did not include the name "Riverine Ecosystem Synthesis." It was only after a series of seminars presented by Jim in Italy, Martin in Australia and South Africa, and intense debate between the three of us that the need for a name of the synthesis and an abbreviation (RES) became evident.

Our initial focus was on fundamental concepts in river ecosystems – an emphasis comparable to almost all lotic models. However, during Jim's trip to Italy, some professors and students were debating which of several prominent lotic models best fit their highly modified rivers. He emphasized at the time that the River Continuum Concept (RCC; Vannote *et al.*, 1980), the Flood Pulse Concept (FPC; Junk *et al.*, 1989), and the Riverine Productivity Model (RPM; Thorp and Delong, 1994, 2002) were all developed for pristine, and now mostly historic, riverine ecosystems. Prior to this European trip, we had considered writing a follow-up paper applying the RES to disturbed environments, but our discussions had never progressed past this speculative stage. At the same time, across the Pacific pond, Martin's work on the structure of riverine ecosystems further developed ideas surrounding functional process zones and their significance to aquatic ecosystems. This work was also starting to be applied to the assessment of the physical character of river networks. Thank goodness for modern technology – the internet and worldwide web – for it facilitated rapid exchanges of different ideas and a growing list of disciplinary-based questions. The chance to expand the RES manuscript into a book finally gave us an opportunity to contribute in this area.

Readers of this book will find that its major emphasis is still fundamental perspectives on the structure and functioning of riverine landscapes – from headwater streams to great rivers and from main channels to floodplains. These perspectives combine aquatic ecology with fluvial geomorphology and landscape ecology. However, two other important components are present. First, we present a recommended guide for applying the theoretical synthesis to actual field analyses. Second, we show how this synthesis relates to riverine landscapes that have been significantly modified in one or more fundamental ways. We believe that it could be vitally important for natural resource managers and for scientists interested in river conservation and rehabilitation take the predictions of the RES into account when developing, for example, monitoring programs encompassing upstream–downstream and channel–slackwater gradients.

Theories should be viewed as formed of unfired clay. They need a lot of shaping and remolding before they accurately model the real world, and sometimes you need to toss them out and start again. Some of the so-called paradigm shifts in environmental science are notable primarily because scientists tend to coalesce for long periods of stasis around popular models rather than constantly committing themselves to the search for truth, as illusive as that goal may be. Although some authors get overly attached to their theoretical models, the big problem is that the users (you the readers and the three of us) too often forget that most models are merely collections of hypotheses no matter the number of disciples that may have jumped upon their *bandwagon*. The problem is aggravated by funding agencies and journals who favor established theories over ideas proposed by nonconformists in their scientific midst. Indeed, some important ideas in ecology have been rejected initially because they seemed too contrary to established ideas or procedures and only later become widely accepted (e.g., Lindeman's trophic dynamic aspect of ecology; Lindeman, 1942; see Sobczak, 2005). In the case of the RES, we have tried to emphasize its heuristic[1] nature whenever we have presented a formal seminar or even discussed the synthesis with professional and student colleagues. We will continue testing the predictions of our synthesis and trying to determine not only where the RES works or does not apply but, more importantly, *why*!

Throughout the book we repeatedly use some abbreviations and somewhat new terms. Here are the principal abbreviations and definitions:

FPC: flood pulse concept
FPZ: functional process zone
HPD: hierarchical patch dynamics (model or paradigm)
IDH: intermediate disturbance hypothesis
PAR: photosynthetically active radiation

---

[1] *Heuristic*: (1) Serving to indicate or point out; stimulating interest as a means of furthering investigation. (2) Encouraging a person to learn, discover, understand, or solve problems on his or her own, as by experimenting, evaluating possible answers or solutions, or by trial and error. Random House Dictionary of the English Language. Second Edition, Unabridged. 1987.

POM: particulate organic matter (CPOM and FPOM = coarse and fine POM, respectively)
RCC: river continuum concept
RES: riverine ecosystem synthesis
RPM: riverine productivity model

Floodscape (an original term): The aquatic and terrestrial components of the riverine landscape that are connected to the riverscape only when the river stage exceeds bankfull (flood stage). These include the terrestrial floodplain (including components of the riparian zone not in the riverscape) and floodplain water bodies, such as floodplain lakes, wetlands, and isolated channels (e.g., oxbows and anabranches).

Functional process zone (FPZ): A fluvial geomorphic unit between a valley and a reach. The name may be a bit confusing to river ecologists because the word *functional* is associated in that scientific discipline with ecological processes, such as system metabolism and nutrient spiraling. However, the term is based on a hydrogeomorphic perspective of rivers, with function being related to dynamic physical processes occurring over time. Moreover, the term FPZ was published prior to our team getting together.

Riverine landscape: The continually or periodically wetted components of a river consisting of the riverscape and the floodscape.

Riverscape: The aquatic and ephemeral terrestrial elements of a river located between the most widely separated banks (commonly referred to as the bankfull channel or active channel) that enclose water below floodstage. These include the main channel, various smaller channels, slackwaters, bars, and ephemeral islands.

In closing, we want to acknowledge the help of many colleagues in developing this book. Although we wrote this entire text, many other people contributed to its success. These include coauthors of some of our previous journal publications (especially Kevin Rogers and Chris James at the University of the Witwatersrand, South Africa, and Melissa Parsons University of Canberra, Australia) and the highly competent people at Academic Press (AP/Elsevier) who helped us produce the book. In the last case, we owe a large measure of gratitude to Andy Richford, who worked with us from the time we first discussed the project with various publishers almost through final production and marketing of the book. We are also grateful to Nancy Maragioglio who developed the original contract for the RES book and sold it to her bosses at Elsevier and to Mara Vos-Sarmiento who led the production effort for this book.

We are also grateful to our students and colleagues at our respective universities who participated in early conversations about the book's content, reviewed material, and/or contributed in other ways. These include Bryan Davies (University of Cape Town, South Africa) Sara Mantovani (University of Ferrara, Italy), Katie Roach (Texas A&M University), students at the University of Kansas (Brian O'Neill, Sarah Schmidt, and Brad Williams), and various Australian contributors, including Scott Rayburg, Michael Reid, and Mark Southwell (University of Canberra), Craig Boys (NSW Fisheries), and Heather McGinness (CSIRO). A big thanks to Renae Palmer for her courage in taking Martin's scribbles and turning them into excellent diagrams.

Finally, we would like to thank the original authors of the River Continuum Concept (Robin Vannote, Wayne Minshall, Ken Cummins, Jim Sedell, and Bert Cushing) and the Flood Pulse Concept (Wolfgang Junk, Peter Bayley, and Rip Sparks) for stimulating many young and older scientists to think conceptually about stream ecology, even though we have disagreed on occasions with these exceptionally good ecologists about aspects of the structure and functioning of riverine ecosystems!

To the readers, we hope you enjoy this book and that it makes you think, even if you disagree with all or parts of it. The number and types of hypotheses included in this book (see Chapter 6 in particular) continue to grow, and we welcome your comments in general along with suggestions for additional model tenets (see section on Concluding Remarks).

Respectfully,
Jim, Martin, and Mike

Acknowledgments

"*To my wife who has stood beside me in*
good times and bad for many wonderful years."
Professor James H. Thorp

"*To Dianne who has been my support pillar*
and reality check for so long."
Professor Martin C. Thoms

"*To Robin and Savannah for their patience*
and support during my academic meanders."
Professor Michael D. Delong

Introduction to the Riverine Ecosystem Synthesis

Background and scope
Basic concepts in the Riverine Ecosystem Synthesis

BACKGROUND AND SCOPE

Conceptual Cohesiveness

Researchers in many scientific disciplines have long sought to integrate their field's diverse theories and models into a small set of core principles, but with only minor successes and much frustration. Within aquatic ecology, Stuart Fisher (1997) wrote a candid and rather damning account of the contributions of the habitat-defined field of *stream* ecology to advancements in the discipline of *general* ecology. His conclusions were based in part on a conviction that concepts in stream ecology were either too habitat-specific or lacked explicit links to general ecological theory. Although the drive within aquatic ecology to develop this conceptual cohesiveness is not comparable in fervor to physicists seeking a unified field theory, attainment of this scientific goal would still be monumental for stream ecology and the contributions of our discipline to ecology in general would be that much greater. However, even if a broad conceptual theory of riverine ecology was judged insignificant by scientists from the perspectives of their own terrestrial and marine habitats, that cohesive theory could substantially advance our own aquatic discipline – especially if it had both theoretical and practical applications – and thus should be a meritorious goal in its own right. Indeed, we maintain that a concurrent goal of all conceptually oriented riverine scientists should be to make their models useful to the large group of environmental scientists and managers who have the difficult task of extracting bits of fundamental theory and applying them to real-world situations. We suspect in most cases that riverine theories are either dismissed as impractical outside of academia or misapplied because of a lack of usable approaches embedded in the theory – the blame for which could be equally laid on the doorsteps of the theorist and the applied scientist.

As in evolutionary biology (Mayr, 1970), a viable cohesive theory is unlikely to be a single factor model and would probably include a balance of conflicting forces. An effective synthesis could draw upon special case theories (e.g., the Flood Pulse Concept (FPC) in lowland floodplain rivers; Junk *et al.*, 1989) but should be more than a compilation of such models. We might be better served in seeking this goal by actively discarding nonviable theories, but the easier (more collegial?) way is usually to let time accomplish that chore. One problem with this approach is that our general textbooks tend to retain old theories long past their prime.

Given that riverine ecosystems are rather mercurial in time and space compared to the average type of global ecosystem, a broad conceptual theory of riverine ecology might contribute more to general ecology by emphasizing this environmental variability. In an analogous fashion, Fisher (1997) suggested that an exploration of ideas about stream shape and its functional consequences could be an opportune area for contributions of stream ecology to general ecological theory.

Our contribution to this quest for a general riverine theory is the heuristic *Riverine Ecosystem Synthesis* (RES; Thorp *et al.*, 2006). As summarized in the current chapter, the RES is an integrated model derived from aspects of other aquatic and terrestrial models proposed from 1980 to 2007, combined with our perspectives on functional process zones (FPZs) and other aspects of riverine biocomplexity. The RES pertains to the entire riverine landscape, which includes both the floodscape and the riverscape. This contrasts with many lotic models whose primary emphasis or support focuses on main channel systems within headwaters. This synthesis, which incorporates the ecosystem consequences of spatiotemporal variability across mostly longitudinal and lateral dimensions, has three broad components:

1. A fundamental, physical model describing the hierarchical patchy arrangement of riverine landscapes within longitudinal and lateral dimensions (Fig. 1.1) based primarily on hydrogeomorphology and emphasizing a new geomorphic division (an FPZ) between the reach and the valley scale (see Fig. 4.3);
2. Ecological implications of the physical model in terms of an expandable set of 17 general to specific (testable) hypotheses, or *model tenets*, on biocomplexity, which is applicable in some form to both pristine and altered riverine landscapes;
3. A framework for studying, managing, and rehabilitating riverine landscapes through the use of the hierarchical physical model and aquatic applications of the terrestrially derived hierarchical patch dynamics (HPD) model (Wu and Loucks, 1975).

Our goal is to provide a framework for the development of a cohesive theory of riverine ecosystems over time rather than to produce a finished product between the covers of this book. In this task, we draw upon three primary components of river science that contribute to the study of riverine landscapes: lotic ecology, landscape ecology, and fluvial geomorphology (Fig. 1.2).

Organization of this Book

The original 2006 publication of the RES in the journal *River Research and Applications* included an historical perspective, a description of the hydrogeomorphic model, application of the HPD model to riverine ecosystems, and development of 14 RES hypotheses (tenets). Our discussion of the RES in this book is divided between two sections (fundamental and applied) and eight chapters. We begin in Section 1 with a broad overview of the RES (Chapter 1), place the RES in a historical context (Chapter 2), and describe hierarchy theory, patch dynamics, and their combination (HPD) in riverine ecosystems (Chapter 3). From there, we explain the importance of a hydrogeomorphic approach for analyzing riverine systems from a theoretical perspective (Chapter 4) and then describe actual methods for defining FPZs for rivers in multiple continents using top-down (e.g., remote sensing) and bottom-up approaches (Chapter 5). This fundamental section ends with an examination of the ecological implications of our hierarchical, hydrogeomorphic model in the form of 17 biocomplexity tenets (Chapter 6), which cover a range of topics from species distributions to landscape processes. Given that pristine rivers in economically developed and developing countries are now almost always part of our lost environmental heritage, we included Chapter 7 in our more applied section to explore both effects of river regulation on model tenets and how the model tenets could be used to manage and rehabilitate riverine landscapes. These perspectives are meant to aid scientists in examining both fundamental and applied aspects of riverine landscapes. Chapter 8 is designed to serve as a roadmap for application of the RES to environmental problems dealing with monitoring, assessment, management, conservation, and rehabilitation of riverine ecosystems.

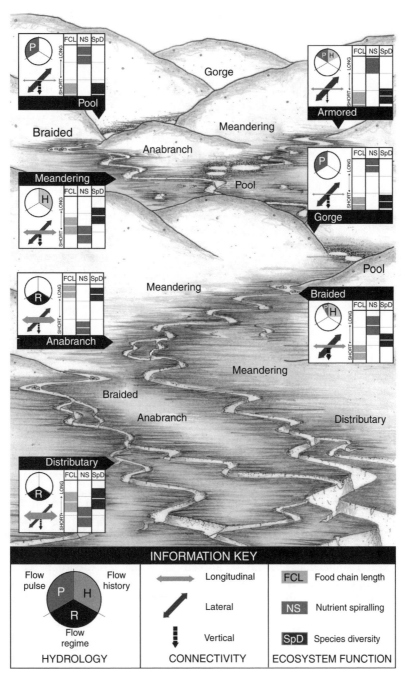

FIGURE 1.1 A conceptual riverine landscape is shown depicting various functional process zones (FPZs) and their possible arrangement in the longitudinal dimension. Not all FPZs and their possible spatial arrangements are shown. Note that FPZs are repeatable and only partially predictable in location. Information contained in the boxes next to each FPZ depicts the hydrologic and ecological conditions predicted for that FPZ. Symbols are explained in the information key. Hydrologic scales are flow regime, flow history, and flow regime as defined by Thoms and Parsons (2002), with the scale of greatest importance indicated for a given FPZ. The ecological measures [food chain length (FCL), nutrient spiraling, and species diversity] are scaled from long to short, with this translated as low to high for species diversity. The light bar within each box is the expected median, with the shading estimating the range of conditions. Size of each arrow reflects the magnitude of vertical, lateral, and longitudinal connectivity (See color plate 1).

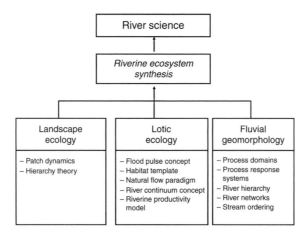

FIGURE 1.2   Contribution of the Riverine Ecosystem Synthesis towards conceptual cohesiveness in the field of river science. The RES specifically brings together concepts and paradigms from the disciplines of landscape ecology, lotic ecology, and fluvial geomorphology.

BASIC CONCEPTS IN THE RIVERINE ECOSYSTEM SYNTHESIS

Many prominent riverine theories, such as the River Continuum Concept (RCC; Vannote *et al.*, 1980) and the FPC (Junk *et al.*, 1989), have included diverse ecological and theoretical elements, but later analyses or applications of these concepts have tended to emphasize only one or two elements – unfortunately, either ignoring the remainder or lumping the entire package (concept) into *agree or disagree* categories, resulting in a less sophisticated and often polarized discussions in the scientific literature. For the RCC, the idea of a longitudinal continuum of species and functional feeding groups based on stream size/order was the focus of most comments in the literature, and aspects of energetics were largely ignored. In the case of the FPC, the rapidly acknowledged importance of the physical flood pulse overwhelmed any serious discussion of the validity of the potentially important trophic model.

Although our abilities to prognosticate are not especially acute, we predict that much of the response to the RES will focus on its physical model (see Fig. 1.1). This is primarily because the implications of adopting this physical perspective on rivers could be far-reaching and secondarily because the HPD component of the RES, while still useful and potentially important, is more complex and somewhat onerous to implement in real-world situations. We hope that individual biocomplexity tenets of the RES will also spur various research projects, but these hypotheses are in themselves probably not bases for major shifts in stream ecology. However, some tenets could be used as benchmarks to test successes or failures in protecting, managing, and rehabilitating riverine ecosystems.

Hydrogeomorphic Patches and Functional Process Zones

Two seemingly simple, but important principles upon which we have based the RES are that riverine ecosystems possess a four-dimensional nature (Ward, 1989) and that rivers are more than a single thread passing through a terrestrial landscape (cf., Ward and Tockner, 2001). The longitudinal and lateral (primarily riparian inputs) nature of rivers was emphasized in the RCC, but Ward (1989) focused attention on vertical (e.g., hyporheic) and temporal dimensions also. The lateral dimension and temporal variability of rivers formed the core of the FPC, and the temporal dimension was emphasized by the natural flow regime paradigm (Poff *et al.*, 1997).

From a longitudinal perspective, most riverine ecologists over the last century have viewed rivers as being either composed of fixed zones (for 80% of the twentieth century) or laid out as a predictable cline, or *continuum*. More recently, a few voices in the wilderness (e.g., Montgomery, 1999; Rice *et al.*, 2001; Poole, 2002; Benda *et al.*, 2004; Thoms, 2006; Thorp *et al.*, 2006) have begun viewing rivers as discontinua rather than as predominately ecological clines.

In contrast to a common view of rivers as continuous, longitudinal gradients in physical conditions, the RES portrays rivers as downstream arrays of hierarchically scaled, hydrogeomorphic patches formed by various factors, principally including catchment and valley geomorphology, hydrologic patterns, riparian conditions, and climate. These patches extend longitudinally, laterally, and vertically, and they may vary temporally from subseasonal to geological time periods. The nature of these hydrogeomorphic patches can differ substantially among and within patch types in spatiotemporal gradients of hydrology and structural complexity. Hydrogeomorphic patches exist at multiple scales, such as drainage basins, valleys, reaches, sets (e.g., a riffle–pool sequence), and units (e.g., a riffle). Missing in this traditional division, however, is a hydrogeomorphic patch intermediate in scale between valleys and reaches. This spatial division has been termed an *FPZ* (Thoms, 2006). The "functional" portion of the name was designated prior to the development of the RES and primarily reflects the perspectives of fluvial geomorphologists. This may be a bit confusing to aquatic ecologists who tend to think of function in terms of ecological processes, such as nutrient spiraling and net ecosystem metabolism. In the RES use of the term, however, it pertains to the physical functioning of geomorphic and hydrologic forces, which shape sections of a riverine ecosystem and thereby alter ecosystem structure and function. An easy, initial way to think of these patches is to envision hydrologic and geomorphic differences among river sections that are either constricted, meandering, braided, or possessed of broad floodplains to cite a few simplistic examples. Although these broad categories of hydrogeomorphic patches are familiar to most river scientists, individual patch types need to emerge from statistical analysis of either top-down [e.g., digital elevation model (DEM) data and remote sensing imagery] or bottom-up approaches using standard techniques employed by fluvial geomorphologists to characterize a site's geomorphic and hydrologic structure (see Chapter 5).

Although previous portraits of riverine landscapes have commonly acknowledged longitudinal differences in channel morphology, these models have generally assumed that the positions of these large hydrogeomorphic patches are consistent and predictable in location and inherently nonrepetitive (otherwise the river would not be a continuum but instead a series of patches). Moreover, the clinal perspective in stream ecology has viewed the essential ecological structure of the community and the nature of ecosystem processes as undergoing a more or less smooth transition from headwaters to the mouth of great rivers. In contrast, the RES contends that many types of FPZs appear repeatedly along the longitudinal dimension of the riverine ecosystems (see Fig. 1.1), with some types of patches repeating more often than others. Although some categories of FPZs are moderately predictable in general position along a riverine ecosystem from higher to lower altitude (see Chapter 5), the topography of the basin and channel and the regional (downstream) differences in climate can obscure these general patterns. Above the ecoregional spatial scale, in fact, the longitudinal distribution of FPZs can be difficult to forecast even for the most experienced fluvial geomorphologist.

Ecological Attributes of Functional Process Zones

Relatively abrupt transitions can exist in the absolute or relative composition of biotic communities and in the nature of ecosystem processes between hydrogeomorphic patches at all scales. Unique ecological conditions exist among types of FPZs because of physicochemical habitat differences, which influence ecosystem structure and function. Different types of hydrogeomorphic patches will vary spatially and temporally in current velocities and patterns of their respective flow pulses (<1 year), histories (1–100 years), and regimes (>100 years). Substrate characteristics, riparian and aquatic vegetation, and water chemistry respond to these climate-related variables as well as to other geomorphic environmental attributes. They, in turn, engender differences in the composition of the biotic community, trophic complexity, and the nature of most ecosystem processes, such as system metabolism, nutrient spiraling, and hyporheic exchange. Consequently, it can be misleading to characterize whole riverine ecosystems as being *floodplain* or *constricted*, for example, because this conveys the false impression that such rivers will consistently differ from each other throughout their lengths and will fit only one of two functional categories.

The process of specifying FPZs is more than an interesting academic exercise because there are significant environmental implications of sampling in different hydrogeomorphic patches. For example, let us assume that a scientist or a manager from a government agency, an environmental company, a nongovernment organization (NGO), or a university has been charged with designing and implementing a sample scheme for monitoring and bioassessing a riverine ecosystem. If this person views the river as a continuum, then the sampling program might be based on a longitudinally stratified, random sample design. In contrast, if the riverine ecosystem is composed of large hydrogeomorphic patches (e.g., our FPZs), as we and others are increasingly proposing, then startlingly different data could be produced depending on the distribution of these patches along the riverine ecosystem and how the samples are distributed among them. Moreover, if the same agency selects reference site by stream order within an ecoregion, they could subsequently sample in highly disparate sites (e.g., constricted and meandering FPZs) with different attributes of community structure and ecosystem function regardless of the state of stream impairment. The identification of FPZs could also contribute to river rehabilitation by enhancing the ratio of ecological services produced by the project compared to the economic costs. For example, the decision to set back levees at a certain distance from the main channel could be based in part on the nature of the FPZ existing in that site before the levees were first constructed.

Hierarchical Patch Dynamics

This RES provides a framework for understanding both broad, often discontinuous patterns along longitudinal and lateral dimensions of riverine ecosystems and local ecological patterns across various temporal and smaller spatial scales. The former is addressed in part by concepts related to hydrogeomorphic patches and FPZs. The latter, as discussed below and in Chapter 3, results from an ecological marriage of patch dynamics and hierarchical classification and is embodied in a set of postulates contained within the HPD model or paradigm (Wu and Loucks, 1995; Wu, 1999), which was originally based on terrestrial systems. This model can be used as a framework for understanding what regulates biocomplexity at various temporal and smaller spatial scales of the riverine ecosystem, such as within a single FPZ (Dollar *et al.*, 2007). Patch dynamics is a concept that was briefly popular in the stream ecology literature during the late 1980s (e.g., Pringle *et al.*, 1988), but it has since then been largely ignored in both theoretical and empirical studies of aquatic systems. Hierarchical classification became popular in the mid-1980s through 1990s (e.g., Frissell *et al.*, 1986; Townsend and Hildrew, 1994; Poff, 1997) and is still discussed in the literature (e.g., Thoms and Parsons, 2002; Parsons *et al.*, 2003; Dollar *et al.*, 2007); however, it is mostly a subject of conceptual papers and has rarely been applied in the field. We believe that both patch dynamics and hierarchical classification procedures are important for stream ecologists to appreciate, but we understand that their practical application can be somewhat challenging, especially with constraints imposed by short grant cycles and restricted funds.

The HPD model integrates a general theory of spatial heterogeneity (patch dynamics) with hierarchy theory by expressing relationships among pattern, process, and scale in a landscape context. It should not be confused with the more restricted concept of nested hierarchical classifications, which was linked to a stream's physical template by Frissell *et al.* (1986). In our original journal article and in Chapter 3, we modify five components of this terrestrial model for riverine ecosystems, which are as follows: (1) nested, discontinuous hierarchies of patch mosaics; (2) ecosystem dynamics as a composite of intra- and interpatch dynamics; (3) linked patterns and processes; (4) dominance of nonequilibrial and stochastic processes; and (5) formation of a quasi-equilibrial, metastable state. Chapter 3 includes a discussion of examples of these five components along with a practical guide for applying this theory in empirical research on riverine ecosystems. For example, we illustrate what kinds of temporal and spatial factors need to be considered for a given research question or environmental problem.

Although the HPD model and this portion of our book may seem irrelevant to applied scientists at first glance, a deeper appreciation of the model will show that some of its components may help explain patterns present in nature and aid in planning strategies for monitoring, managing, and rehabilitating riverine landscapes at the correct hierarchical scales of time and space.

Bicomplexity Tenets

Many testable hypotheses can be generated from the RES, but manuscript space restricted the number included in Chapter 6 and in our original journal article. We have limited these hypotheses both to the functioning of epigean portions of riverine ecosystems and to ecological timescales. They are focused more on the riverscape than on the entire riverine landscape (*sensu* Wiens, 2002), an emphasis partially reflecting the more recent developments in river–floodplain research. We attempted to integrate previous discussions of FPZs and HPD with these tenets, but the HPD model is less commonly incorporated, in part because it is more of a scientific approach than an environmental descriptor. We make no claim to originality for all these tenets. Some of these ideas are well supported in the scientific literature, whereas others may be controversial or border on being educated guesses. We have expanded the number of tenets from 14 in our journal article (Thorp *et al.*, 2006) to 17 in this book, and we have modified most of them slightly or substantially to make them more applicable across a broad range of river types (encompassing a larger geographic perspective) and to incorporate ideas from our unrelenting exploration of the nature of riverine ecosystems.

The first set of tenets (1–4) concerns factors influencing species distributions or, in effect, composition of the species pool. They emphasize the importance to species distributions of patches of different sizes, the nature of the FPZ (as opposed to location on a longitudinal dimension), ecological nodes (e.g., transitions between and within FPZs), and current velocity vs hydrologic retention.

The next section on community regulation relates to factors controlling species diversity, abundance, and trophic complexity within the assemblage of species potentially present in the environment; both density-independent and density-dependent factors are included. Tenets 5–9 relate to the importance of the habitat template, deterministic and stochastic factors, quasi-equilibrial conditions, and various types of ecological succession within different sections of the riverine landscape.

The final set of tenets (10–17) covers processes at the ecosystem and riverine landscape levels. Among topics covered are differences in primary productivity among FPZs, algal grazer vs decomposer food pathways throughout the riverine landscape, effects on the relative importance of organic energy sources from a community's location along both longitudinal and lateral dimension of the riverine landscape, nutrient spiraling and FPZs, the importance of the natural flow regime, the relationships among flood-linked life histories and flood seasonality, effects on ecosystem processes of aquatic connectivity, and influence of spatial arrangement of FPZs on biocomplexity.

Historical and Recent Perspectives on Riverine Concepts

INTRODUCTION

Understanding the ecological structure and function of natural or altered riverine ecosystems is a common goal of many stream and river ecologists. This has spurred the development of numerous conceptual models, shaped empirical research and funding, and occasionally altered government policies on river management and rehabilitation. Formation of conceptual theories can expand our knowledge of factors regulating river networks as long as popular theories are viewed as the "latest best approximations" rather than iron-clad truths and if ecologists seek to test theories and comprehend why concordance or incongruity emerge.

A single chapter in this book is insufficient to explore the nature and applications of all prominent riverine theories published even in the last few decades. Consequently, we are focusing our review and analysis on only those hypotheses, models, theories, and paradigms that address (i) large-scale spatial patterns affecting the structure and function of riverine ecosystems and (ii) ecological regulation of communities at smaller spatiotemporal scales. At the larger spatial scale, we concentrate our analysis on two (longitudinal and lateral) of the four recognized dimensions of rivers (Ward, 1989). Although a third, vertical dimension (e.g., exchange with the hyporheic zone) is important to ecosystem functioning, we only briefly cover it here because less controversy seems to exist among stream ecologists about processes and patterns operating in this dimension. The fourth dimension, which involves temporal phenomena, is treated in multiple contexts throughout this book. By the longitudinal dimension, we are alluding to patterns and processes occurring along discharge and altitudinal gradients from headwaters downstream to the river mouth. And by the lateral dimension, we are referring to similarities and differences in communities from the main channel through slackwaters (riverscape) to the floodplains (floodscape). At smaller spatial scales, we discuss theories debating which biotic and/or abiotic factors regulate community structure and the importance of temporal phenomena.

We hope that a more thorough understanding of historical differences and commonalties in riverine models will serve as a starting point for developing a conceptual consensus in riverine ecology (see Chapter 1). Because this chapter is also meant to set the stage for discussion of the RES, our review of selective aspects of other models is tailored to that synthesis and its specific contribution toward conceptual cohesiveness. Our failure to discuss or fully analyze all models is not a judgment of their usefulness in general but instead reflects primarily their utility in explaining the nature and applications of the RES.

PATTERNS ALONG A LONGITUDINAL DIMENSION IN RIVER NETWORKS

For more than a century, stream ecologists have been interested in differences in aquatic communities from headwater streams to large rivers. Researchers have addressed many aspects of longitudinal processes (i.e., upstream–downstream changes), such as concepts involving energy sources and allocation (Vannote et al., 1980; Thorp et al., 2006), nutrient spiraling (Newbold et al., 1982), river network and landscape interactions (e.g., Montgomery, 1999; Gomi et al., 2002; Benda et al., 2004), and serial discontinuity and dams (Ward and Stanford, 1983b). Early attempts to cope with this complexity by dividing riverine ecosystems into specific, longitudinally ordered zones (e.g., Hawkes, 1975) were widely accepted and are still applied in some countries (Santoul et al., 2005). This approach, however, later came under strong criticism (e.g., Townsend, 1996), especially after publication of the RCC (Vannote et al., 1980), one of the most influential riverine papers of the twentieth century. The RCC portrayed riverine systems as intergrading, linear networks from headwaters to the mouths of great rivers. It is currently the dominant theory employed intentionally or *de facto* by riverine ecologists and environmental scientists/managers. It is also taught in classrooms throughout the world and represents one of the very few acknowledged contributions of stream ecology to general ecology.

Influential perspectives on longitudinal patterns published prior to the current century are discussed below. The concept of large hydrogeomorphic patches (FPZs), along with related concepts, is briefly discussed here, but is the focus of Chapters 4 and 5.

Longitudinally Ordered Zonation

Initial attempts to divide rivers into biotic zones appeared in European scientific literature in the latter part of the nineteenth century and early part of the twentieth century (reviewed in Hynes, 1970; Hawkes, 1975). Zones were initially defined based on which of four dominant fish species characterized the fauna: trout (*Salmo*), grayling (*Thymallus*), barbel (*Barbus*), and bream (*Abramis*). The overall fish fauna associated with these indicator species were also described along with the geomorphic nature of the streams providing the characteristic habitat. Proponents claimed that the zones could be predicted from the knowledge of stream width, bed slope, and valley shape (e.g., Huet, 1954 in Hawkes, 1975). This initial work on streams in Germany, Belgium, and Poland was modified by Carpenter (1928) for British streams and then by many other authors for riverine systems of other regions and continents, such as in North America by many proponents (e.g., Kuehne, 1962). Hynes (1970) described limitations in using zones based primarily on fish species whose distributions vary among geographic regions and which may be absent for natural historic reasons or because of human activities. Nonetheless, he partially supported this approach, which was used extensively around the world for more than 75% of the twentieth century and which occasionally is still employed by some fish biologists in Europe and elsewhere (e.g., Santoul et al., 2005). In addition to zones based primarily on fish distribution, many other authors described longitudinal zonation patterns for autotrophs (algae, mosses, and vascular plants) and benthic invertebrates.

Biotic names for different zones rapidly proliferated in both lentic and lotic systems and threatened to overwhelm the search for conceptual and functional understanding of river networks in a flood tide of verbiage. In response, riverine ecologists began adopting more general, habitat descriptions to define river zones. In a series of papers, Illies (e.g., 1961 in Hynes, 1970) proposed a generalized scheme for classifying rivers around the world. He had observed marked faunal changes from the lower limit of the presumed grayling zone and the upper range of the barbel zone, which also coincided with major shifts in water temperature. He called these the rhithron and the potamon, respectively, and divided each into epi-, meta-, and hypo- rhithron or potamon sections. To the rhithron was later added the eucrenon and the hypocrenon (also called the krenal) in recognition of the presence of springs and spring brook regions, respectively (Illies and Botosaneanu, 1963). Streams derived from the meltwater of glaciers and permanent snowfields were termed kryal by Steffan (1971). Although these longitudinal names were proliferating only slightly, additional names for lateral components began appearing in response to greater attention to aquatic floodplain ecology; these include eu-, plesio-, para-, and paleopotamal (e.g., Ward *et al.*, 1999).

Several problems are encountered when designating *fixed* zones in rivers, especially when defined by biotic communities rather than by the basic hydrogeomorphic character of a river network. As was obvious from the earlier attempts in this area, major geographic and ecoregional differences exist globally, which have restricted the number of useful definitions that rely primarily on distributional differences in biota. This approach has also been criticized (Townsend, 1996) for its implications of a Clementsian-like (Clements, 1916) nature of aquatic communities as tightly coevolved entities somewhat akin to superorganisms. Fixed zones, whether biotically or abiotically bounded, are incompatible with the concept that a single zone might appear multiple times within a river and in different locations among rivers. Typically the zones are named and applied based on only a few features (e.g., biota, temperature or a few aspects of geomorphology). Finally, as underscored by criticisms in the RCC, one could infer that fixed zones are ecologically isolated from each other rather than components of a stream continuum. This last shortcoming spurred the development of the RCC.

THE RIVER AS A CONTINUUM – A CLINAL PERSPECTIVE

The concept of rivers as consisting of strongly ordered longitudinal zones was largely and rather rapidly abandoned at the beginning of the penultimate decade of the twentieth century as a direct consequence of the publication of a single paper – the RCC (Vannote *et al.*, 1980) – along with subsequent elaborations and modifications to the original model (e.g., Minshall *et al.*, 1983, 1985 and Sedell *et al.*, 1989). Central to the RCC were the linked concepts that (i) physical variables within a river network present a *continuous gradient* of physical conditions from headwaters to a river's mouth and (ii) this longitudinal gradient "... should elicit a series of responses within the constituent populations resulting in a continuum of biotic adjustments and consistent patterns of loading, transport, utilization, and storage of organic matter along the length of a river" (Vannote *et al.*, 1980). For simplicity sake, the model assumed an uninterrupted gradient of physical conditions in natural rivers where physical conditions gradually alter as one moves downstream. Minshall *et al.* (1983) noted, however, that, "... regional and local deviations [*from RCC predictions*] occur as a result of variations in the influence of (1) watershed climate and geology, (2) riparian conditions, (3) tributaries, and (4) location-specific lithology and geomorphology." The patch-forming effects of tributaries and dams on RCC predictions for pristine systems were also discussed in the serial discontinuity concept (e.g., Ward and Stanford, 1983b). Note that these deviations from the continuum nature of river networks were considered *exceptions* and not threats to the fundamental portrayal of rivers. In contrast, other authors (e.g., Poole, 2002; Thorp *et al.*, 2006, and see next section) contend that these "exceptions" are in fact the rule.

The 1980 RCC paper included a surprisingly wide diversity of ideas, but aquatic ecologists have focused on only the two central themes related to longitudinal progression of organic food sources and functional feeding groups from headwaters to great rivers. From perspectives on how physical conditions should theoretically alter the relative and absolute input of allochthonous carbon and generation of autochthonous organic matter, the RCC postulated a predictable, unidirectional change in functional feeding groups from small streams to large rivers. Although the concept of functional feeding groups itself has been criticized (e.g., Mihuc, 1997), these categories and their relationship to a predicted continuum were the subject of many research studies in the 1980s and are still widely cited in general biology and ecology textbooks.

Every theory has its critics, some with valid points, and the RCC is no exception. Until recently, however, few people have challenged the basic continuum perspective; they merely clashed on other aspects of the model, such as the RCC's original portrayal of headwaters as being forested and the consequent implications for food webs (e.g., criticisms based on New Zealand streams in Winterbourn *et al.*, 1981). In the RCC's defense, the original paper included a caveat acknowledging this ecoregional focus, but the point was deemphasized by the authors and largely ignored by readers thereafter. As noted above, conflicts were treated as exceptions to the general rule by the RCC authors (e.g., Minshall *et al.*, 1985) rather than as basic challenges to fundamental properties of the model. One early and rare criticism of this continuum assumption came from papers by Statzner and Higler (1985, 1986). They contended that stream hydraulics were the most important environmental factor governing zonation of stream benthos on a worldwide scale. Rather than a steady gradient of stream hydraulics postulated by continuum models, they identified discontinuities where transition zones in flow and resulting substrate size were the critical determinants of changes in species assemblages.

The most serious, early disagreement with the RCC came with the publication of the FPC (Junk *et al.*, 1989; Junk and Wantzen, 2004); however, this model did not argue against a fundamental principle of the RCC that river networks consist of a continuous gradient of physical conditions from headwaters to a river's mouth. Instead, they took issue with predictions that large river food webs were based energetically on organic matter derived from upstream processing inefficiencies and argued instead for the primacy of floodplain processes. Following the publication of the FPC, the RCC was modified in the same symposium proceedings by some of the original RCC authors (Sedell *et al.*, 1989). They concluded that the RCC required revision for the floodplain portions of large rivers but was still entirely appropriate for constricted channel rivers. The applicability of the RCC to dryland rivers was first criticized by Walker *et al.* (1995), and this has continued with research by other dryland ecologists (Bunn *et al.*, 2006; Thoms, 2006). Other ecologists have disagreed with some trophic predictions of the model for channels of large rivers (see review in Thorp and Delong, 2002 and brief coverage in Chapter 6) or have even concluded that the idealized downstream pattern of the RCC in primary trophic resources "... is remarkable primarily because it is not usually realized and cannot provide a worldwide generalization" (Townsend, 1989).

A perhaps minor point in the original RCC was the hypothesis that the biotic structure and function of a stream community conforms to the *mean state* of the physical system over time. The essential nature of the RCC is not dependent on the validity of this hypothesis, which was framed on the basis of the energy equilibrium theory of fluvial geomorphology. However, a large and persuasive body of evidence accumulated in the last two decades suggests that environmental variability is at least as important in shaping biotic communities as the mean state. Most research in this area has focused on the role of flood and flow pulses (e.g., Poff *et al.*, 1997; Tockner *et al.*, 2000), but droughts play a very important role in some systems (Boulton, 2003; Lake, 2003; Carroll and Thorp, unpublished data). Both forms of variability could play a greater role in some ecoregions in determining community structure and ecosystem function than the mean state of the environment. This topic is discussed further in Chapter 4 in terms of the flow patterns uniquely characterizing each type of geomorphic FPZ and their associated ecological FPZ.

Hydrogeomorphic Patches vs a Continuous Riverine Cline

The two previous perspectives on community changes along a longitudinal dimension in riverine landscapes (fixed biotic zones and a stream continuum) together almost completely dominated the entire twentieth century in stream ecology. A fixed zonal perspective is no longer intimately associated with a particular set of authors, but a clinal or continuum perspective, for better or worse, is still closely linked with the RCC in the minds of perhaps most aquatic ecologists. This is in part because the term *continuum* is used both in the general sense of a riverine ecosystem and in the specific sense of the theory embodied in the RCC. In a similar way, the ecological importance of floods (or a flood pulse) is closely linked to the groundbreaking and influential FPC (Junk *et al.*, 1989). It is too easy for nonspecialists, and students in particular, to confuse general, noncontroversial terms (continuum and flood pulse) with specific predictions of individual theories, which may or may not be subject to controversy.

Serious challenges to a clinal perspective did not appear until the new millennium (other than Statzner and Higler, 1985, 1986). The basis of the dissent was that this portrayal does not work physically because it underestimates the importance of differential geology within a catchment, tributary effects, and historical geomorphic influences, and merely considers these to be exceptions to the continuum rule. Contrasting models on longitudinal changes in river networks are briefly discussed below and are treated more intensely in Chapter 4. These new models, including some aspects of the RES, agree with the RCC that some *predictable* changes in habitat, community structure, and ecosystem function occur along a longitudinal dimension in river networks. For example, at a large spatial scale, the particle size of benthic sediments gradually diminishes as you move from headwaters to large rivers, at least in most systems with high-relief headwaters. Likewise, the size and the recalcitrant nature of suspended particulate organic matter (POM) gradually decrease or increase, respectively, as you move downstream. Moreover, distributions of many higher taxonomic groups vary predictably at *very large* scales from headwaters to large rivers. For example, species richness of fish generally rises with mean river discharge. Newer models differ, however, in at least three major areas, which are the following: (i) the relationships among large adjacent reaches (i.e., a mostly gradual cline vs a disjunct pattern); (ii) the relative importance of longitudinal position vs local condition (e.g., an FPZ); and (iii) the degree of predictability among rivers and ecoregions in community structure and ecosystem processes at any given point downstream.

Some critics of a clinal perspective have argued that a predictable downstream pattern may exist from a large-scale perspective but it is *not* characterized by gradual biotic adjustments or consistent patterns of loading, transport, use, and storage of organic matter. For example, Statzner and Higler (1986) disagreed with the concept that steady changes in stream biota existed along a purported continuum in riverine ecosystems; instead, they felt that changes were abrupt and occurred at hydrologic and substrate transition points in the riverine ecosystem. In their "Link Discontinuity Concept," Rice *et al.*, (2001) noted the lack of recognition of the effects of tributaries in the original RCC (but see Minshall *et al.*, 1985) and proposed that at a moderate spatial scale (1–100 km), changes in substrate particle size typically follow a punctuated, sawtooth pattern highly susceptible to tributary influences and strongly affecting the longitudinal distribution of macroinvertebrates. In a similar vein, Benda *et al.* (2004) identified tributary junctions with the main channel as biological hot spots in their "Network Dynamics Hypothesis." Townsend (1996) also concluded that the nature of changes from upstream to downstream in vertical and lateral connectivity is influenced heavily by the stream segment structure and tributary catchments. He proposed a broad spatiotemporal framework termed the "Catchment Hierarchy."

Other opponents or modifiers of a clinal perspective have maintained that longitudinal patterns are less predictable than proclaimed by the RCC and have observed that changes are more reflective of local conditions than the position along the continuum. Montgomery (1999) concluded that a clinal perspective was valid only for low-relief watersheds with relatively constant climate and simple geology, whereas his "Process Domains Concept," based on the importance of local geomorphic conditions and landscape disturbances, was applicable in regions with high relief, variable climates, and complex geology (e.g., the U.S. Pacific Northwest). Walters *et al.* (2003) concurred with this assessment for the Etowah River of northern Georgia, USA. Stanford and Ward (1993) sought to modify the RCC for alluvial rivers by describing the longitudinal, *beaded* series of aggraded alluvium and linked ecotonal processes that produce predictable groundwater communities and other aquifer-riverine convergence properties along the continuum from head-waters to large rivers. Perry and Schaeffer (1987) found only a weak downstream gradient in benthic species in a tributary of the Gunnison River in Colorado and no gradient in functional guild composition; they characterized species distributions as punctuated gradients.

Poole (2002) departed more substantially from previous conceptual analyses and proposed that rivers are composed of patchy discontinua where the community responds primarily to local features of the fluvial landscape. A stream's discontinuum, according to Poole, "... is comprised of a longitudinal series of alternating stream segments with different geomorphic structures. Each confluence in the stream network further punctuates the discontinuum ...." He noted that changes in the branching pattern of a riverine ecosystem and variation in the arrangement of component patches (roughly comparable to the RES' large hydrogeomorphic patches) along a downstream profile can result in substantial changes in predicted pattern of solute concentration and create gaps in the downstream transitions in community structure (see also Rice *et al.*, 2001). Consequently, Poole concluded that a biotic community within a stream segment is not necessarily more similar structurally and functionally to communities in adjacent segments than it is to assemblages located farther upstream or downstream – a view that he contrasts with a clinal perspective. He hypothesized that the degree of divergence from a clinal pattern was influenced by the location of the discontinuity along a longitudinal profile of the riverine ecosystem.

Our present book and the manuscript on which it has been primarily based (Thorp *et al.*, 2006) emphasize a nonclinal view of riverine ecosystems but still acknowledge the presence of some very large-scale changes in ecosystem structure and function along a longitudinal dimension. We agree with the conclusion of Poole (2002) that local hydrologic and geomorphic conditions are more important to ecosystem structure and function than simple location along a longitudinal dimension of the riverine ecosystem. We also concur that adjacent FPZs (our terminology) can be less similar than pairs separated by greater distances. In that sense, we conclude that downstream patterns are much more dissociated from upstream processes than once thought. Finally, the RES stipulates that FPZs can be replicated within a riverine ecosystem, and their placement within and among riverine ecosystems – while certainly not random – is also not fixed along a longitudinal dimension (see Fig. 1.1). Indeed, one's ability to predict the location of a given type of patch diminishes with increasing geographic scale, especially above the ecoregional level.

Although the riverine concepts emerging around the turn of the century (e.g., Montgomery, 1999, Rice *et al.*, 2001, Poole *et al.*, 2002, Thorp *et al.*, 2006) diverged substantially from a continuum approach, it is important to avoid confusing modern nonclinal perspectives with the earlier fixed zonation viewpoint of nature. Both recognize fluvial geomorphic influences, but the earlier models (i) stipulated fixed community properties – even naming them for organisms; (ii) failed to explicitly recognize large-scale longitudinal patterns other than zonal; and (iii) did not consider repeatable zones, variable zonal order, or effects of zonal position on its inherent characteristics.

Network Theory and the Structure of Riverine Ecosystems

Rather than analyzing riverine ecosystems as downstream arrays of patches along a longitudinal dimension or as a simple river continuum, rivers can be evaluated using fractals (Mandelbrot, 1982; Veitzer *et al.*, 2003), multifractals (De Bartolo *et al.*, 2000), and network approaches. Network theory is a relatively recent development by statistical physicists (Albert and Barabási, 2002; Newman, 2003), which is being used to examine many types of patterns in nature, including biochemical properties, human social interactions, food webs (e.g., Krause *et al.*, 2003), lake invasion routes for exotics (Muirhead and MacIssac, 2005), and the transport of sediments in (Tayfur and Guldal, 2006) and human colonizers along river networks (Campos *et al.*, 2006). Network theory, as applied to linkages among tributaries in a downstream progression, focuses on rivers as nodes and tributary confluences as links and considers them as having a scale-free architecture. In their Network Dynamics Hypothesis, Benda *et al.* (2004) concluded that deviations from the expected mean state of conditions within a channel occur in response to network geometry and that the tributary junctions serve as ecological hot spots. Network theory does not emphasize basic changes to the hydrogeomorphic nature of the river other than spatially limited increases in habitat complexity.

The application of network theory to riverine ecosystems is in its infancy, but many potential applications having a spatial context may be revealed in the next decade. Although its use for explaining biotic communities and ecosystem processes *within* a given area is probably limited, it may prove useful in examining processes occurring *among* river locations. For example, there are potential applications to the spread of invasive species, energy and nutrient flow, and metapopulation dynamics of fish.

THE LATERAL DIMENSION OF RIVERS – THE RIVERINE LANDSCAPE

Imagine describing an iceberg as a large piece of ice floating on the open ocean surface – such an incomplete description is somewhat comparable to describing a river as a single channel cutting through the surrounding watershed or portraying it as a blue line on a white wall map. It might seem trite to say to aquatic scientists that rivers are not limited to their main channels nor to the visible surface waters; but, for various reasons, that is the way rivers were viewed for many years from both ecological and management perspectives. Perhaps more important is that this is the common perception of nonscientists, including those holding the purse strings for research and management support, as it is only recently that a complete and more accurate portrayal of the riverine landscape has begun to emerge in the scientific community.

Until the role of floodplains was emphasized in the FPC by Junk *et al.* (1989), rivers were often erroneously discussed as being a single channel of flowing water, much like a thread passing through a terrestrial landscape (cf., Ward and Tockner, 2001). Since the landmark (floodmark?) FPC publication, the definition of rivers has improved but is still often inaccurately applied. Many aquatic ecologists and environmental managers tacitly treat rivers as if their lateral dimension consisted only of the main channel and its supra-bankfull floodplains. Ignored in this categorization are the ecologically vital slackwaters – sub-bankfull regions either continuously or frequently connected to the main channel, though sometimes with an absence of detectable surface currents (Thorp and Casper, 2002). They include many shorelines, shallow to deeply incised bays, secondary and side channels, alluvial wetlands, and backwaters (where rising waters back up into semi-enclosed areas lacking upstream connections except during some flood periods). These hydrologic retention areas correspond closely to the terms "dead zone" and "storage zone" often used by European scientists (see also Hein *et al.*, 2005). We agree with Tockner *et al.* (2000) that the hydrologic flow pulse may be as important ecologically in these sub-bankfull areas as the flood pulse has proved to be in supra-bankfull, floodplain areas.

Rather than being a single channel of flowing water, riverine ecosystems are composed laterally of the riverscape (sensu Wiens, 2002) and the floodscape, which are hierarchically nested within the riverine landscape by both structural and functional properties. The complexity of both riverscape and floodscape varies with the type of FPZ. The main channel and various slackwater areas below the geomorphic bankfull mark (the traditional riverine border of the floodplains) constitute the riverscape proper. Also included in the riverscape are ephemeral bars and islands, which are periodically submerged by flow pulses. Thus, the riverscape includes the lower portion of the aquatic/terrestrial transition zones, or ATTZ, as defined by Junk *et al.* (1989). The amount of the riverscape that is covered by water below flood stage varies spatially and temporally by ecoregion and FPZ according to local precipitation patterns. In very arid ecoregions, such as much of Australia, the riverscape may contain no surface water or perhaps only isolated pools for periods lasting from weeks to more than a year. The *floodscape* (our original term) consists of (i) a planform known as floodplains, which is demarcated by the extent of alluvial sediments; (ii) relatively permanent (i.e., past the flood pulse) floodplain lakes, ephemeral ponds, many oxbows (billabongs), and periodically disconnected anabranches (which can also be a part of the riverscape) located within alluvial sediments; and (iii) bare or vegetated areas of the floodplains that are primarily aquatic only when inundated by a flood pulse (this constitutes the remainder of the ATTZ). Other appropriate but more complex approaches to characterizing river landscapes were described by Ward *et al.* (2002).

Structural features and functional processes of the riverscape differ substantially from those found in aquatic and terrestrial components of the floodscape (see discussion in Chapter 6). Moreover, if the entire riverscape is considered – and certainly if the more encompassing riverine landscape is assessed – our perspectives on longitudinal patterns in biodiversity and ecosystem processes could be altered. For example, would the common assumption that biodiversity peaks in mid-order streams/rivers be valid if the lateral components of large rivers were factored in, rather than comparing only main channel reaches?

The effects of these implicit, misleading definitions of rivers (albeit often unintended) have been reinforced inadvertently by an overwhelming research emphasis on low-order streams by lotic ecologists, the linear nature of influential clinal theories such as the RCC and the nutrient spiraling theory (Newbold *et al.*, 1982), a paucity of empirical research on structurally complex floodplain rivers, and artificial channelization of many floodplain rivers (Thoms and Sheldon, 2000). As Fisher (1997) pointed out, the major paradigms and research foci in stream ecology over the past quarter century "... have been based upon a linear ideogram – an image which is at best incomplete and at worst, incorrect." Furthermore, as long as we persist in labeling the main channel as lotic and the side, low- to zero-flow slackwaters in the riverscape as semilentic, we risk misunderstanding roles played by these two interdependent components of riverine landscapes, which may be equally integral to functioning of streams and rivers.

Although attention to the ecology of floodplain rivers has blossomed profusely since the publication of the FPC, most of this research has dealt with spatial components of the riverine landscape that are seasonally dry or at least isolated from the main channel via surface waters. For example, intriguing studies of the Danube River (e.g., Hein *et al.*, 2003) have shown dramatic ecological effects on community structure and ecosystem processes associated with temporal length of connectivity (short periods of aquatic connectivity of floodplain areas with the Danube main channel interspersed with long spans of surface isolation). Such research is helping to promote rehabilitation of large European rivers (e.g., Hein *et al.*, 2005). Likewise, changes in the lateral organization of zoobenthos in floodplains have been examined along the longitudinal dimension of rivers (Arscott *et al.*, 2005). In contrast, very few conceptual or empirical studies that focus on patterns and processes in lateral areas of the riverscape, i.e., the moderate- to zero-flow slackwater areas (but see Ward *et al.*, 1999, 2002; Tockner *et al.*, 2000), have been undertaken. Questions abound on the relationship between hydrologic connectivity – measured, for example, as the time it takes for a molecule of water from the main channel to reach any given point in the slackwaters – and both patterns and processes related to biodiversity, food web complexity, nutrient spiraling, and system metabolism. The dearth of research on lateral components of rivers in general may have resulted from several factors including the historical and modern, dominant focus of riverine ecologists on headwaters systems (vs rivers, especially large rivers) and the sizeable loss of lateral areas of rivers from the construction of levees and various bank stabilization projects.

Another fundamental property of the lateral nature of rivers relates the temporal component to spatial heterogeneity, which is related to flow variability. This has received very little consideration by riverine scientists, but is briefly discussed in the following section.

TEMPORAL DIMENSION: NORMALITY OR ABERRATION?

The philosopher Heraclitus (*ca.* 535–475 B.C.) is often quoted as describing the impossibility of stepping in the same river twice. If one blithely ignores this Greek's penchant for riddling and obscurity and then accepts his wording literally, we could recognize him as an ancestral stream ecologist who appreciated the variability of riverine ecosystems – and incidentally who scooped modern fluvial geomorphologists! In another sense, however, environmental scientists, conservation biologists, and river managers need to acknowledge that both periodic and aperiodic changes in a river's flow at various spatiotemporal scales do *not* alter its fundamental ecological nature. That is, variability in flow and, to some extent, structural complexity are inherent features of all rivers, just like the marine intertidal would not be the same ecosystem without the temporal patterns of the tides. In that sense, therefore, Heraclitus could be considered misinformed ecologically because one can step into the same, but normally fluctuating, river twice! In summary, *riverine changes are normal, while constancy is usually an aberration.* Ignoring this simple principle is one of the more significant causes of environmental problems in rivers throughout the world.

Philosophical considerations aside, scientists have begun changing the portrayal of floods and droughts as aberrations and have been replacing them over the last two decades with a characterization of rivers as temporally variable in both flow and spatial complexity. Probably the publication that stirred the most interest in this area among river scientists was the natural flow regime paradigm of Poff *et al.* (1997), which linked flow variability with management, conservation, and rehabilitation of rivers. However, this important paper was based on a foundation of previous studies by many fluvial geomorphologists, such as Leopold *et al.* (1964), Schumm (1977a), and Walker *et al.* (1995). Although many factors are clearly involved in controlling pattern and process in riverine ecosystems, flow regime is thought to be the master control variable (Power *et al.*, 1995; Poff, 1997; Resh *et al.*, 1998). Or as described by Walker *et al.* (1995), flow is the "maestro that orchestrates pattern and process" in riverine ecosystems (Thoms, 2006).

Precipitation events in the watershed lead to a pulse of water through the ecosystem, which is termed a *flow pulse* if it directly affects only the riverscape and a *flood pulse* if it spreads across the floodscape (Fig. 2.1). In a flood pulse, the river tops its banks, creating a "flood" in both the traditional and the legal sense. In the latter case, the river bank height, watershed

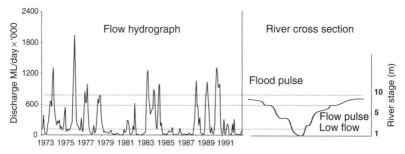

FIGURE 2.1   River hydrograph and river channel cross section illustrating low flow, flow pulse, and flood pulse for the Darling River at Bourke, southeastern Australia.

topography, and historical flow patterns are used in many countries to predict the spatial extent of damaging flood waters with mean recurrence intervals of usually 100 or 500 years. Flood pulses are characterized by relatively large lateral movements of waters, whereas flow pulses include fluctuations of water only within the riverscape. Flood pulses spread water to many areas where long-lasting or frequent submergence would destroy the extant botanical assemblages. Exceptions are areas such as the Amazon basin where highly predictable and long-lasting floods occur every year, allowing evolutionary adaptations to flooded riverine landscapes. In contrast, plant assemblages in the riverscape are either adapted to occasional submergence of stems or include botanical species capable of rapid recruitment following a drop in river stage. The bottom of the wetted riverscape lacks true soil and is composed of fine to coarse sediments and rocks, whereas the floodscape is composed of alluvium and soil derived from local plant decomposition. The term *hydrologic connectivity* in the floodscape describes the frequency of river surface waters either entering an area of previously dry land or merging with lentic waters on the floodscape, while connectivity in the riverscape is a parameter that can be defined as the time it takes for a water molecule or dissolved/suspended particle to reach any given point in the slackwaters from the river's main channel. Many of the ecological processes resulting from flow and flood pulses are similar (e.g., regeneration of nutrients, dispersal of propagules, and redistribution of sediments), but in other cases the effects are different because organisms in riverscapes and floodscapes, respectively, have adapted to different habitat conditions and constraints.

A river's natural flow regime includes both temporal and spatial components of the habitat, which in turn alter physical, chemical, and biotic patterns and processes throughout the ecosystem. Most empirical studies in this area have focused on floods because droughts are primarily limited to intermittent headwater streams around the world and to low–high stream "orders" in arid and semiarid environments – both environments have, unfortunately, attracted the interest of few aquatic ecologists. Research has progressed furthest in arid regions, such as Australia (e.g., Lake, 2000; Thoms, 2003; Bunn *et al.*, 2006; Sheldon and Thoms, 2006a), South Africa (Dollar *et al.*, 2007), and in both the Great Plains (e.g., Dodds *et al.*, 2004) and southeastern region of the USA (e.g., Schade *et al.*, 2001; Dahm *et al.*, 2003). An important result of these studies is an appreciation of the role of drought periodicity to ecosystem patterns and processes. Scientists in the modern grant-funded environment are accustomed to thinking of flow or drought effects manifested in a single year (flow/flood pulses and possibly annual droughts), but longer-term flow histories (1–100 years) and regimes (>100 years) can be crucial to explaining the presence/absence of species and levels of ecosystem processes (Thoms and Sheldon, 2002; Thoms and Parsons, 2002, 2003). Flow variability differs among stream sizes, ecoregions, and even continents. For example, in a comparison of streamflow gauges among five continents, Poff *et al.* (2006) determined that gauges throughout the contiguous United States exhibited the greatest mean, overall flow variation but those in the mostly arid Australian continent showed the largest interannual variability. The rising human demand for water and the likelihood of increased stream intermittency in arid and semiarid ecoregions, such as the U.S. Great Plains (Dodds *et al.*, 2004), may spur increased government funding for drought-related research and perhaps even better integration of scientists, stakeholders, and service agencies (Rogers, 2006) concerned with droughts and floods.

The temporal nature of the presence/absence of water and its movement (advection and hydraulic effects) in riverine ecosystems has multiple impacts on physical, chemical, and biotic patterns and processes. Hydraulic impacts of water currents in general, and pulsed events in particular, alter many aspects of the physical environment in the riverine landscape. At a very small spatial scale, the size of sediment particles transported along the bed or in suspension may change during a river pulse, and the mere movement of the sediment can disturb or kill adults of small species or the larvae or other propagules of many taxa. At a larger spatial scale, the abundance and the size of hydrologic retention areas are affected by pulses, and the types, sizes, and spatial arrangement of landscape patches below and above the water surface can be significantly transformed. Moreover, a direct association exists between flow variability and channel complexity at the cross-sectional scale, with channel complexity having a strong effect on ecosystem integrity (Thoms, 2006). River pulses bring nutrients into channel margins, into more isolated backwaters of the riverscape, and into lentic habitats of the floodscape during critical periods. Although much of the riverscape seems well supplied with nutrients, concentrations of nitrogen and phosphorus in isolated backwaters fluctuate widely, potentially limiting primary production (Knowlton and Jones, 1997; Tockner *et al.*, 1999a; Richardson *et al.*, 2004). This can produce shifts in autotrophic species (e.g., from green algae to N-fixing cyanobacteria; e.g., Huff, 1986; Knowlton and Jones, 1997), which in turn can alter the efficiency of energy transferred among trophic levels.

The nature of the ecological responses to hydrologic pulses is intimately linked to their timing and predictability. Although short-term pulses can dramatically alter physicochemical conditions in the riverine landscape, life history adaptations often require longer-term and more predictable pulses. It is also ecologically important in many ecoregions to have these pulses associated with warmer temperatures to match conditions promoting higher primary productivity with periods of maximum potential for metazoan growth and reproduction in newly submerged areas of the riverscape or floodscape. Although access to flooded areas is not necessarily essential for most fish species in some rivers, such as temperate rivers with aseasonal flood cycles, it usually greatly enhances overall fish recruitment (Humphries *et al.*, 1999, 2002; Winemiller, 2005).

VERTICAL DIMENSION: THE BULK OF THE ICEBERG!

Although the lateral dimension is increasingly recognized as an integral component of rivers, the vertical dimension is often ignored when defining riverine boundaries, despite Hynes's 1983 call for better incorporation of groundwater studies into stream ecosystem concepts. This is partly the "out of sight, out of mind" syndrome and partially the result of having few riverine scientists studying groundwater systems. Early groundwater studies were mostly limited to karst ecosystems, but Gibert *et al.*'s 1994 book on groundwater ecology included analyses of hydrogeomorphology, biological organization, ecosystem processes, human impacts, and surface–subterranean interactions in both karst systems and the porous media of unconsolidated rocks. Aquatic ecological studies of unconsolidated media have concentrated more on the hyporheic zone (areas partially influenced by epigean water currents) than the phreatic zone (an area farther from the stream and intergrading with the hyporheic zone; this area is relatively uninfluenced by surface-water movement).

As in surface waters, the hyporheic zone can be represented as a hierarchy of geomorphic patches (Poole *et al.*, 2006) interacting hydrologically in a dynamic fashion with both epigean and phreatic zones. Stanford and Ward (1993) proposed the Hyporheic Corridor Concept to describe ecotonal processes in alluvial rivers along large spatial scales in longitudinal and lateral dimensions. According to this model, serial patches of aggraded alluvium in the floodplains alternate with constricted regions of the river to produce a landscape pattern resembling "beads on a string."

Groundwater flow is relatively stable compared to that in epigean zones but is much more dynamic than previously thought (Wondzell and Swanson, 1996, 1999; Malard *et al.*, 1999). Its dynamic nature is influenced by the composition, density, and spatial arrangement of the unconsolidated sediments and rocks within and above the hyporheic zone, and thus should vary with the type of FPZ in the riverine ecosystem. Current velocity and hydrologic retention fluctuate in surface waters and groundwaters in response to altered spatial complexity of epigean and hyporheic zones and are significantly influenced by both increased (flow and flood pulses) and decreased river discharge (zero-current flow or complete loss of surface water). These changes in downwelling of epigean waters and upwelling of groundwaters alter flow pathways and their organic and inorganic signatures (e.g., Dent *et al.*, 2001; Malard *et al.*, 2001; Malcolm *et al.*, 2003), creating a patchy and dynamic habitat mosaic (Poole *et al.*, 2006) of abiotic conditions for organisms in both the hyporheic zone and aquatic components of the surface riverscape and floodscape (e.g., Boulton and Stanley, 1995; Boulton *et al.*, 2002; Stanford *et al.*, 2005).

Most geomorphic modifications of rivers undertaken to regulate flow or floods (e.g., dams and levees) should alter spatial complexity and water movement in the surface and subsurface components of the riverscape and floodscape, with complex effects on flow pathways along vertical and lateral dimensions. These in turn should affect the associated biotic communities and ecosystem processes, such as nutrient spiraling (e.g., LeFebvre *et al.*, 2004). Hydrogeomorphic drivers of groundwater flow paths are discussed in recent papers by Poole *et al.* (2006) and Stanford (2006).

OTHER IMPORTANT RIVERINE CONCEPTS

The aquatic theories discussed above have concentrated on concepts related in some way to ecological aspects of hydrogeomorphology, and have thus emphasized the four dimensions of riverine ecosystems. As we indicated at the beginning of this chapter, there are many other models that have been important to aquatic ecology, which we intentionally omitted in this book because of their minor applications to the RES and/or space limitations. However, concepts related to hierarchy theory, patch dynamics, and system equilibrium are discussed in Chapter 3.

Hierarchical Patch Dynamics in Riverine Landscapes

HIERARCHICAL PATCH DYNAMICS MODEL – BRIEF INTRODUCTION

Understanding the nature of changes in biocomplexity from headwaters to a river mouth is an important path toward developing a conceptually cohesive model of riverine ecosystem structure and function. This large-scale perspective is not meant to fully explain the regulation of biocomplexity at various temporal and smaller spatial scales of the riverine ecosystem, such as within a single FPZ. For that purpose, we turn now to another major component of our model – aquatic applications of the HPD model devised by Wu and Loucks (1995; see also Wu, 1999). The original, terrestrial-based HPD model integrates a general theory of spatial heterogeneity (patch dynamics) with hierarchy theory by expressing relationships among pattern, process, and scale in a landscape context. Although they share a few common features, this HPD model should not be confused with the more restricted concept of nested hierarchical classifications, which was linked to a stream's physical template by Frissell *et al.* (1986). Relevant aspects of this concept are summarized below, and the model is described in detail by Wu and Loucks (1995) as an important ecological paradigm. Following this brief review of the HPD, we summarize hierarchy theory and patch dynamics before discussing the application of the HPD model to riverine ecosystems.

In our analysis of HPD, we first need to define what we mean by a *patch*. This task is harder than it might sound to the layman because the size of a patch is scale-, organismal-, and process-dependent and can vary greatly in temporal dimension and size (e.g., an individual rock to a river segment or a floodscape area). Furthermore, species of different sizes, life histories, and evolutionary traits will often experience physical hierarchies and patches from contrasting perspectives (cf. Hildrew and Giller, 1993), and the discreteness of the habitat boundary hinges upon the organism's motility (Tokeshi, 1993). Likewise, patches from a species perspective are typically scaled differently from those from a process perspective. Perhaps the best, albeit far from satisfactory, way to describe a patch is as *a spatial unit differing from its reference background in nature and appearance*, a depiction that could also be applied to temporal patches. This definition does not constrain the size or internal homogeneity of the patch and is rather loose in its requirements for discreteness. In the case of FPZs, the FPZ is a large hydrogeomorphic patch that is smaller than a valley but larger than a reach. It can be delineated statistically using top-down (e.g., remote imagery) and bottom-up approaches with common techniques in fluvial geomorphology (see Chapter 5). However, most scientists do not work at the FPZ level and are thus interested in smaller patches, perhaps down to the microscopic level. These sub-FPZ patches need to be defined by the investigator for the species and process examined. An important caveat to keep in mind is that the scientific interpretation of patterns and processes is highly influenced by the spatial and temporal scale of the patch definition (e.g., Thompson and Townsend, 2005a), even if the investigator does not explicitly describe the study as focusing on patches. This can sometimes be responsible for disagreements in the scientific literature about the significance of different processes in nature, such as the role of deterministic and stochastic factors.

The HPD model is composed of five principal elements (modified below from Wu and Loucks, 1995). First, ecological systems are viewed as 'nested, discontinuous hierarchies of patch mosaics.' This allows one to analyze the role of small patches (e.g., substrate types) within large patches (e.g., a riffle or reach). It also enables investigators to incorporate both seasonal and aperiodic changes in the nature and role of patches. The presence of patches within a hierarchy of regulatory factors reflects the action of different disturbances and other independent variables operating over multiple spatiotemporal scales. Second, the dynamics of ecological systems are derived from a composite of intra- and interpatch dynamics. This interaction among patches produces emergent properties of riverine ecosystems, which is not evident when studying patches in isolation. For example, from a study of individual cobbles in a stream, one might conclude that interference competition controls species diversity. But when a large number of cobble stones (patches) are examined, the investigator might decide that entirely different processes are important, such as stochastic stream flow. Third, pattern and process are interlinked and scale-dependent. Various processes (e.g., nutrient spiraling) may create, modify, or eliminate patterns at certain spatial and temporal scales, while at the same time certain spatial and temporal patterns (e.g., differences in flow characteristics) can substantially alter ecological processes. Scale-dependent interrelationships can change from a riffle-sized patch to a channel-floodscape patch and, therefore, may require different approaches to elucidate. For example, if one asked what controls community diversity within a riffle during the summer, conducting resource limitation and predator–prey experiments might prove fruitful. In contrast, understanding ecosystem functional responses to variability in flow patterns might be a more profitable approach at the FPZ scale within a river with a broad floodscape. Fourth, nonequilibrial conditions and stochastic processes play a dominant role in the so-called 'ecosystem stability.' Deterministic processes can still contribute significantly to community regulation within a given patch; but on a hierarchical scale, stochastic processes among patches are more important, as discussed in model tenet 6 in Chapter 6. Fifth, a quasi-equilibrial, metastable state can develop at one hierarchical level through incorporation of multiple, nonequilibrial patches from the adjacent, lower level – in essence, 'out of chaos comes order!' (See model tenet 7 in Chapter 6.)

HIERARCHY THEORY

The spatiotemporal complexity of river ecosystems requires modeling approaches that can handle a high level of variation in multiple river dimensions. Unfortunately, most river ecosystem concepts and studies tend to be locked into or work from a descriptive base with a strong emphasis on river classification or a description of pattern and process, or they model a restricted set of attributes of rivers. There are many issues surrounding cross-disciplinary approaches to the study of ecosystems. Apart from the traditional approach of individual disciplines attempting to understand their own system and then adding extra relationships specific to the study at hand, Walters and Korman (1999) suggested that the interaction between disciplines is often conducted at inappropriate scales. For example, current biological monitoring techniques, such as AusRivas (Norris and Hawkins, 2000) and Rivpacs (Wright, 2000), use a series of large-scale catchment variables to predict the macro-invertebrate communities, which may occur at a small-site scale. In these examples, no consideration is given to how these large-scale variables may be related to ecological processes, or whether variables operating at other scales may explain an equivalent amount of biological variation.

Studies of river systems are often designed to test hypotheses through the traditional scientific method of falsification. This approach is frequently achievable and has been successful in advancing our knowledge of rivers. However, the prerequisite for this approach is that there should be simple cause-and-effect relationships between the factors under consideration. River ecosystems are characterized by physical features and organisms with both individual histories and interactions that form multicausal relationships. The study of river ecosystems needs to deal with multicausal relationships at different scales and across different disciplines. The classical approach to science from individual disciplines may then be inappropriate for a full understanding of river ecosystems and may, in fact, inhibit this goal. Indeed, Pickett *et al.* (1999) suggested that the new philosophy of science should be scale-sensitive and move away from the conventional reductionist falsification approach, which limits the development of an appropriate understanding of complex systems such as rivers. This demands a hierarchically based approach that integrates description, causal explanation, testing, and prediction of riverine ecosystems (Pickett *et al.*, 1999).

River scientists from many disciplines commonly attempt to organize research problems to take into account some attributes of spatiotemporal scale. Geomorphologists have long recognized the importance of time, space, and causality. In their landmark paper, Schumm and Lichty (1965) provided a functional typology of dependent, independent, and indeterminate variables. They concluded that at a reach scale, channel morphology should exhibit a long-term average state, which would depend on variables like discharge, vegetation, and sedimentology. They also contended that smaller spatiotemporal variables, such as turbulence, should be irrelevant in these cases. Similar examples can be found in hydrology and biology (Barrett *et al.*, 1997; Thoms and Parsons, 2002). These studies recognize the hierarchical nature of physical systems and the need to identify appropriate scales in order to establish rigorous cause-and-effect relationships. River ecosystems have multiple hierarchies – in this case a geomorphological, hydrological, and freshwater ecology hierarchy, each interacting with the other (Fig. 3.1).

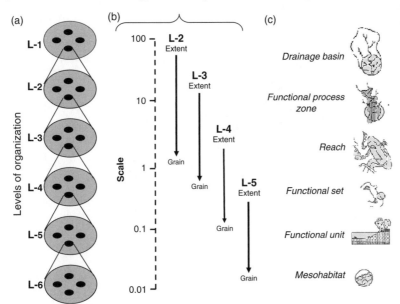

FIGURE 3.1 Hierarchical organization of patches within a riverine landscape. (a) A conceptual diagram of hierarchical patches where patches at one scale are nested within a level of organization above; (b) the scale associated with a hierarchically organized system and the implications of grain and extent based on hierarchy theory; and (c) physical patches of the riverine landscape that can be recognized at various levels of organization and scale.

Hierarchy is common to hydrology, geomorphology, and freshwater ecology and may be essential for establishing interdisciplinary studies of riverine ecosystems. Viewing riverine ecosystems from a truly interdisciplinary perspective requires the establishment of linkages between the hierarchies of different disciplines (cf. Thoms and Parson, 2002; Thorp *et al.*, 2006; Dollar *et al.*, 2007; Parsons and Thoms, 2007).

Rivers are natural hierarchical ecosystems that can be resolved into different levels of organization (Werner, 1999) (Fig. 3.2). A level, or holon, is a discrete unit of the level above and an agglomeration of discrete units from the level below. Separate levels can be distinguished by frequencies or rates that differ by one or more orders of magnitude (O'Neill and King, 1998), and subsystems with similar frequencies or rates occupy the same level within a hierarchical system. Higher hierarchical levels have more sluggish process rates or frequencies and, therefore, react more slowly than lower levels. Levels of organization and scale are not equivalent. The former is essentially a relative ordering of systems (see Fig. 3.1), and units for designating levels are absent. Consequently, the common, interchangeable use of the term level and scale is inappropriate (King, 1997). Although a level of organization is not a scale, *per se*, it can have a scale or can be characterized by scale (O'Neill and King, 1998). Scales can be both spatial and temporal, but the former is more commonly applied. Scale refers to the dimensions of observed phenomena and entities. It is recorded as a quantity and involves measurement and their units. These measurement units are then used to characterize and distinguish between objects or the frequency of processes. Scale is used to assign or identify dimensions and units of measurement and answer the question of how big is a catchment, riverine ecosystem, etc. This can be stated only with a scale; hence, spatial scale is the physical dimension of an inanimate or even living object. This term can also refer to the scale of observation, that is, the spatial and temporal dimensions at which phenomena are observed. Two aspects of scale are especially important: grain and extent. Grain refers to the smallest spatial or temporal interval in an observation set (O'Neill and King,

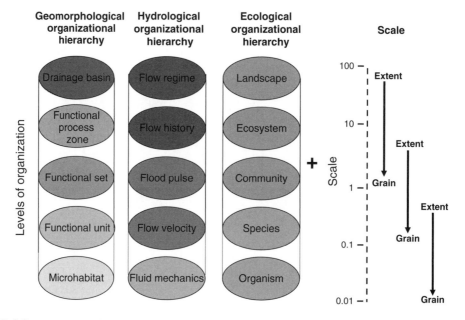

FIGURE 3.2 Organizational hierarchies in river science. To use this framework, one must first define the relevant spatiotemporal dimension for the study or question. Scales for each hierarchy are then determined and allow the appropriate levels of organization to be linked. The scale at the right demonstrates that linking levels across the three hierarchies may be vertical depending on the nature of the question (See color plate 2).

1998), for example, the smallest scale in a pattern to which an organism will respond. Extent is the total area or length of time over which observations of a grain are made or the largest pattern to which an organism may respond, for instance, a fish's home range.

River scientists from all disciplines commonly organize problems according to time and space perspectives. Disciplinary paradigms, or just well-established traditional biases, strongly influence the style and scale at which practitioners generally view structure and function within a riverine ecosystems. For example, within freshwater ecology are distinct levels of biological organization, which often correspond to individuals, populations, meta-populations, communities, and ecosystems. Although these are not scales (Petersen and Parker, 1998), they operate in characteristic spatial and temporal domains and are used to stratify components within the biological system. For example, physiology and behavior are generally studied at the level of the individual, whereas diversity is usually studied at the community level.

Geomorphology also organizes river systems in a hierarchical manner. The approach of Frissell *et al.* (1986) is typical and illustrates interrelationships between geomorphological factors at five levels: stream systems, segment systems, reach systems, pool–riffle systems, and microhabitat systems (Table 3.1). Each level of this geomorphological system develops and

TABLE 3.1   Hierarchical Systems Used in the Stream Classification Scheme of Frissell *et al.* (1986)

System level	Definition	Linear spatial scale (m)	Timescale of persistence (years)	Defining Physical Factors
Stream system	All surface waters in a watershed	10^3	10^6–10^5	*Watershed factors:* biogeoclimatic region, geology, topography, soils, climate, biota, culture. *Stream system factors:* watershed class, long-profile slope and shape, network structure
Segment system	Portion of stream system flowing through a single bedrock type and bounded by tributary junctions or major waterfalls	10^2	10^4–10^3	Stream class, channel floor lithology, channel floor slope, position in drainage network, valley side slopes, potential climax vegetation, soil associations
Reach system	Length of a stream segment lying between breaks in channel slope, local side slopes, valley floor width, riparian vegetation, and bank material	10^1	10^2–10^1	Segment class, bedrock relief and slope, morphogenic structure or process, channel pattern, local side slopes and floodplain, bank composition, riparian vegetation state
Pool/riffle system	Subsystem of a reach with characteristic bed topography, water surface slope, depth and velocity patterns	10^0	10^1–10^0	Reach class, bed topography, water surface slope, morphogenic structure or process, substrates immovable in less than 10-year flood, bank configuration
Microhabitat system	Patches within pool/riffle systems that have relatively homogeneous substrate types, water depth and velocity	10^{-1}	10^0–10^{-1}	Pool/riffle class, underlying substrate, overlying substrate, water depth and velocity, overhanging cover

Each system persists over a characteristic spatial and temporal scale and is defined by a characteristic set of physical factors.

persists at a characteristic spatial and temporal scale. Smaller-scale features develop within the constraints set by larger-scale features of which they are a part. The spatial and temporal scales associated with each level subsequently translate into a set of defining physical factors that can be used to identify hierarchical boundaries of each level within a catchment. For example, at the top of the hierarchy, stream systems within a watershed persist at large spatial scales and long timescales and are defined partly by independent catchment factors such as geology and climate. This pattern of characteristic scales of persistence and physical factors continues through the hierarchy of segment, reach, and pool/riffle systems until at the bottom of the hierarchy, microhabitats persist at small temporal and spatial scales and are defined by dependent factors such as substrate, water velocity, and depth. Thus, the division of a catchment into component hierarchical levels can provide a practical representation of the complex interrelationships that exist between physical and geomorphological factors across different spatial and temporal scales.

In hydrology, five levels of hydrological behavior have been identified as being important for river ecosystem functioning (Thoms and Sheldon, 2000). Different levels in the hydrological hierarchy are commonly characterized by a scale, and these are as follows:

- *Flow regime* (long-term, statistical generalization of flow behavior or climate – macroscale influences that extend over hundreds of years and are relevant to continental landmass, catchments, and river channels);
- *Flow history* (the sequence of floods or droughts – mesoscale influences between 1 and 100 years that extend to river channels, zones, and channel cross sections);
- *Flood pulse* (an individual flood event – microscale influences that generally extend less than 1 year and are often related to channel cross sections, bedforms, and boundary sediment composition);
- *Channel hydraulics* (velocity and turbulent fluctuations in three dimensions; bed and shear stresses – nanoscale influences of minutes and seconds that may influence bedforms and boundary-sediment composition);
- *Fluid mechanics* (surface pressures – pica scale influences are likely to have a greater influence on chemical processes than those that are more physical in nature).

In hierarchically organized systems, levels or holons are not completely isolated or strictly independent. Successive levels act like filters and/or constraints. Any level higher in the hierarchy exerts some constraints on all lower levels (O'Neill, 1989), particularly the level immediately below. Boundaries between levels in a hierarchy are generally identified on the basis of functional process rates (O'Neill and King, 1998) or structural spatial criteria (O'Neill *et al.*, 1986). The exchange of information between levels in the hierarchy of organization has three main properties, which are as follows:

1. Specific levels of organization are linked to specific spatial and temporal scales, so that higher levels are linked to broader spatial and temporal scales.
2. Rate differences of at least one order of magnitude exist between different levels, so that higher levels have lower frequencies of behavior than do lower levels.
3. Higher levels constrain lower levels through their larger time constraints (Bergkamp, 1995).

A characteristic feature of hierarchical systems is that higher levels occur at large scales and have slow rates of behavior while lower levels occur at small scales and react more quickly (O'Neill and King, 1998). As such, hierarchies are considered to be *nearly decomposable* because each level of organization responds at a characteristic spatial and temporal scale (Bergkamp, 1995). In addition to being nearly decomposable, hierarchical systems also have emergent properties. Emergent properties are those properties of higher levels that cannot be deduced from the functioning of their parts (Allen and Starr, 1982) and arise because it is only the averaged, filtered, or smoothed properties of a lower level that input to higher levels of the hierarchy (O'Neill *et al.*, 1986). With an increase in the number of intervening levels separating levels of interest, there is a corresponding decrease in the influence of the rapid behavior of a lower level on any level that is above it in the hierarchy (O'Neill *et al.*, 1986; Kotliar and Wiens, 1990). Lower levels in a hierarchically organized system are constrained by conditions imposed by successively higher levels.

In a geomorphological hierarchy, higher level variables set the conditions within which others can form (Schumm, 1988; de Boer, 1992). Geology and climate are independent catchment factors, because they directly or indirectly influence the formation of all other factors in the cascade (Schumm and Lichty, 1965; Lotspeich, 1980; Knighton, 1984; Naiman *et al.*, 1992; Montgomery, 1999). Being at a higher level, climate and geology control catchment physiography as well as the vegetation and soils present in that area – and thus, indirectly, the uses to which humans put the land. In turn, catchment physiography, vegetation, soils, and land use regulate channel-forming factors such as stream discharge, sediment input, sediment transport, sediment caliber, bank characteristics, and valley slope. These channel-forming factors subsequently control the channel geometry and flow factors occurring at lower levels of the river hierarchy, such as channel slope, water velocity, planform and cross-sectional channel dimensions, and the arrangement of bedforms in the channel. Thus, in a fluvial system, geomorphological factors operating at one level of the hierarchy constrain the formation of factors at successively lower levels (Schumm, 1991).

Overlain across this hierarchy is a temporal context that determines the strength of constraint among successive hierarchical levels (Schumm and Lichty, 1965). The relationship between factors located at different levels is a function of the persistence of each factor over time (Table 3.2). In general, the strength of influence between factors located at the top and bottom of the geomorphological hierarchy decreases with an increase in time span. The independent catchment factors, climate and geology, are at the top of the hierarchy, because

TABLE 3.2 The Status of Riverine Landscape Variables During Time Spans of Different Duration

	Status of variables during designated time spans		
Riverine landscape variables	Geologic (millions of years)	Modern (thousand years)	Present (one year)
Time	Independent	Not relevant	Not relevant
Geology (lithology and structure)	Independent	Independent	Independent
Climate	Independent	Independent	Independent
Vegetation (type and density)	Dependent	Independent	Independent
Relief	Dependent	Independent	Independent
Palaeohydrology (long-term discharge of water and sediment)	Dependent	Independent	Independent
Valley dimension (width, depth, and slope)	Dependent	Independent	Independent
Mean discharge of water and sediment	Indeterminate	Independent	Independent
Channel morphology (width, depth, slope, shape, and pattern)	Indeterminate	Dependent	Independent
Observed discharge of water and sediment	Indeterminate	Indeterminate	Dependent
Observed flow characteristics (depth, velocity, turbulence, etc.)	Indeterminate	Indeterminate	Dependent

After Schumm and Lichty (1965).

over a geologic time span of millions of years, they constrain the formation of factors such as vegetation, relief, palaeohydrology, and valley dimension. However, at this time span, factors at intermediate and lower levels within the hierarchy, such as sediment discharge, water discharge, channel morphology, and flow character, are not influenced by any of the higher-level factors (Schumm, 1988). Over a time span of thousands of years, channel morphology becomes the only factor that is influenced by factors operating at higher levels of the hierarchy. For the same reason, lower-level factors, such as sediment discharge, water discharge, and flow character, are unresponsive. This pattern of constraint continues until at a time span of 1 year, sediment discharge, water discharge, and flow character are the only factors influenced by higher-level factors on an annual basis (Schumm and Lichty, 1965). Thus, interrelationships between any of the geomorphic factors in a river system are directly linked to the evolution and behavior of the system at different temporal scales.

Lower levels of organization can also influence the structure and functioning of those at higher levels, and this is dependent upon the nature of the boundary between individual levels in the hierarchy. Boundaries based on gradients in rates are said to show loose coupling between successive levels in a hierarchy (O'Neill *et al.*, 1986). Activity rates within a level and between successive levels alter dynamics of the next higher level in the system. In this way, the structure and function of a lower level can influence the structure and function of the next higher level. In a study of the Arolla River, Switzerland, Lane and Richards (1997) demonstrated the long-term behavior of medial bar forms to be a product of events acting on smaller spatial and temporal scales.

Most hierarchy theories concern a single hierarchy, but riverine ecosystems have multiple hierarchies. Identifying appropriate levels of organization and, therefore, scales between different hierarchies in complex systems has rarely been attempted because of entrenched paradigms within the individual disciplines. Riverine ecosystems cannot be arbitrarily defined in space and time, as is commonly done by individual disciplines. Rather, they must be defined relative to the level of the problem being addressed, and defining and isolating the relevant level in a hierarchy is a critical step in setting up any problem (O'Neill *et al.*, 1986). The different hierarchies present in river ecosystems, as defined by the different disciplines of geomorphology, hydrology, and ecology, also have different levels of organization and associated scales (see Fig. 3.2). In any hierarchy, with a change in observational scale, you may eventually move across a discontinuity in scale thereby changing levels of organization, grain, and extent. This may reduce the power of explanation of cause and effect. This can also occur if you move between different hierarchies because levels of organization and scale may be incompatible between the different hierarchies. Linking levels of organization in different hierarchies can be achieved by matching scales. Riverine ecosystems are scale-sensitive, multiple hierarchies. For each level of a particular hierarchy, appropriate matching variables will be present within the primary hierarchy and the other hierarchies in the ecosystem.

Problem solving in practice is typically reduced to simplistic top–down or bottom–up approaches, thereby limiting and fragmenting solutions especially when dealing with multiple hierarchical systems (Walters and Korman, 1999). The typical approach of individual disciplines in an interdisciplinary setting is to understand your own individual system and then add extra relationships peculiar to your study or issues at hand. To overcome this, Walters (1986) proposed a *working outward* model for interdisciplinary issues that involved identification of key variables and their level within a hierarchy followed by explicit definition of scale. Identifying the appropriate scale enables cross-linking between hierarchies. This approach does not constrain solutions within a linear framework, and it allows key variables along with their respective levels and scale to be linked across hierarchies.

PATCH DYNAMICS DEFINED

Earlier we gave a rather loose definition of the term patch, and the definition of *patch dynamics* is even more vague, because strictly speaking, it merely implies a change in the nature or distribution of a patch (cf. White and Pickett, 1985). Although these definitions are simple, both have profound implications for the structure and functioning of riverine ecosystems. Patches can be studied in terms of their size, shape, composition, relative availability in space and time, and distribution in relation to other patches and environmental features. Patch dynamics adds a crucial temporal component to the provision and use of patches. Patch dynamics is most often studied in relation to biotic and abiotic, disturbance-mediated coexistence of species (e.g., Denslow, 1985; Downes, 1990; Palmer *et al.*, 2000). In nearly all cases involving biota, studies have examined the response of species to the temporal dynamics of patches, especially as related to hydrologically induced changes (e.g., Bowen *et al.*, 2003; Arrington *et al.*, 2005), but in some cases the reverse approach has been adopted (e.g., Statzner *et al.*, 2003). The latter focus is related to many studies of bioturbation in marine and freshwater systems (e.g., Mermillod-Blondin *et al.*, 2004). Most studies of 'patch dynamics' are of simple patches over various spatiotemporal scales. Much less riverine research has been undertaken on interpatch relationships, such as the distribution and arrangement of patches in space and time (but see Palmer *et al.*, 2000).

HIERARCHICAL PATCH DYNAMICS IN RIVERINE RESEARCH

In their original HPD manuscript, Wu and Loucks (1995) identified five elements of HPD, as summarized earlier in the present chapter. After brief discussions of the scales of our analyses and the nature of patch studies in riverine ecosystems, we analyze in greater detail how these elements apply to, and are interpreted for, dynamic aquatic ecosystems.

Selective Spatiotemporal Scales

A simultaneous advantage and disadvantage of the HPD model is its lack of specificity about the temporal or spatial scales and where analyses must start. Frissell *et al.* (1986), in their hierarchical framework for stream habitat classification, listed events and processes controlling stream habitats on different spatiotemporal scales (see Table 3.1). At stream and segment scales, these included geomorphic and hydrologic processes operating through periods ranging from centuries to over 100,000 years. Substantial alterations of stream channels and watersheds, resulting from major climatic shifts, tectonic processes, glaciation, etc., can occur over such lengthy periods. During the same or longer period, evolutionary and long-term ecological processes can interact with abiotic and biotic environmental characteristics to produce the species pool from which riverine communities are drawn. Linking the appropriate spatiotemporal scales with their relevant ecological processes (ecosystem to individuals) may be critical to understanding natural functioning of rivers (Thoms and Parsons, 2002), managing them properly, and rehabilitating them where appropriate.

We have arbitrarily confined our present analysis to ecological time frames, which can be vaguely defined as being shorter than both evolutionary and long-term geological processes. We selected this scale for both brevity of the book and because this scale is most relevant to the most common empirical and theoretical analyses of community structure and ecosystem processes. The spatial scales discussed below are inclusive of the full longitudinal dimension of riverine ecosystems (headwaters to the river mouth) and the entire riverine landscape (main channel to floodscape). They also encompass the watershed level down to microhabitat patches. In this spatial approach, we have a greater chance of understanding differences in the importance of various regulatory factors from headwater streams to great rivers and in applying this information to management and rehabilitation of rivers.

The Nature of Patches and Their Study in Riverine Landscapes

River scientists often implicitly include spatial patches as a research variable in study designs, but relatively few have considered the temporal dynamics of patches until recently. The patches most commonly studied have rarely exceeded dimensions of reaches, riffles, or pools, and most were no larger than individual cobbles or distinguishable resource patches (cf. Woodward and Hildrew, 2002). Initial research on small patches focused on biotic interactions, principally interference competition for space or exploitative competition and often included aspects of predation/herbivory (e.g., Peckarsky, 1979; Georgian and Thorp, 1992) or succession (reviewed in Fisher, 1983). This was followed in the late 1980s and 1990s by a greater appreciation of the role of flow disturbance in regulating species within patches (e.g., Statzner and Higler, 1986) in concert with the broader ecological trend recognizing the importance of stochastic factors in nonequilibrial environments. There now exists widespread agreement in aquatic ecology that organisms are greatly influenced by interactions between physical habitat and flow patterns, the latter being considered the master control variable by many scientists (Resh *et al.*, 1988; Power *et al.*, 1995; Walker *et al.*, 1995; Poff, 1997). These observations promoted the application of the habitat template model to streams (Frissell *et al.*, 1986; Townsend and Hildrew, 1994). Because temporal scales of biotic interactions (e.g., competitive exclusion from small to large patches up to a stream network) are usually different from those related to disturbances from flow variability, a temporal hierarchy of regulatory factors develops, even within small spatial patches (cf. Fisher *et al.*, 1998). The result is a mosaic of dynamic patches in small streams, which vary in composition, size, and recovery stage (quasi-successional?).

Research on large-scale patches has been less common in the field of riverine ecology. However, the RCC, which was written for a generalized riverine ecosystem, implicitly recognized segment-scale patches that were central to its interpretation of species composition, resource availability, and functional feeding groups along its proposed river continuum. These larger-scale patches included: (i) segments at a low stream order with a canopy cover from riparian trees (e.g., headwater streams in forested ecoregions) or without a riparian canopy (e.g., grasslands) and (ii) patches where light penetrated to the bottom in open streams (small to medium rivers) or failed to reach it because of depth and/or turbidity (medium to large rivers). For simplicity sake, the RCC assumed an uninterrupted gradient of physical conditions in natural rivers where physical conditions gradually alter as one moves downstream. From the knowledge of how physical conditions alter the input or generation of allochthonous and autochthonous organic matter, the RCC postulated a predictable, unidirectional change in functional feeding groups from small streams to large rivers. Other authors modified the RCC for the patch-forming effects of tributaries and dams (e.g., Ward and Stanford, 1983b).

On the basis of the theory of conservation of energy, as used in fluvial geomorphology, Vannote *et al.* (1980) hypothesized that the biotic structure and function of a stream community conforms over time to the mean state of the physical system (within these larger-scale patches). This theory was developed before stream ecologists began focusing on the importance of variability in riverine ecosystems. Consequently, the temporal nature of these patches, and possible importance of variable physical states, was not factored into the model.

With increased study of large rivers, the potential importance of large lateral patches (e.g., floodplains and large slackwaters) has become evident. The FPC (Junk *et al.*, 1989) emphasized a significant lateral component in extensive floodplains. Trophic dynamics within these FPZs were shown to be distinctly different and divorced from upstream areas. This contrasted with the prevailing idea that they were linked by fine particulate organic matter (FPOM) derived as a consequence of upstream processing inefficiencies. In recognition of this lateral component, Sedell *et al.* (1989) indicated that the RCC is still applicable to upstream reaches and large constricted-channel rivers but is an inappropriate model for floodplain rivers. At a different spatial scale compared to those in large rivers, small streams can also have moderate-sized (e.g., riffles and pools) through large patches (e.g., beaver dam pools; e.g., Naiman, 1997).

Publication of the FPC was of great help in expanding the formerly small trend of research in large rivers (e.g., Winemiller, 2004). Scientists began recognizing the importance of channel complexity around the world (e.g., Walker *et al.*, 1995; Thoms and Sheldon, 1997, 2006; Amoros and Bornette, 2002; Robinson *et al.*, 2002) and initiated studies comparing ecological structure and functioning among and within large patches (FPZs) formed by channel braiding, anabranches, various forms of slackwaters, floodplains, etc. Often these large patches were created, eliminated, or altered extensively on seasonal or aperiodic temporal scales by sub-bankfull flow pulses (Tockner *et al.*, 2000; Burgherr *et al.*, 2002; Thoms *et al.*, 2005) or supra-bankfull flood pulses (Junk *et al.*, 1989). Recognition of the ecological importance of differences in flow velocity, turbulence, and retention time in large rivers led to the development of inshore retention concept (Schiemer *et al.*, 2001a). A growing number of scientists are now finding that large lateral, flow-related patches in large rivers are responsible for major differences in structural and functional attributes of riverine ecosystems for plankton, benthic macroinverte-brates, nutrient cycling, productivity, and trophic food webs (e.g., Vranovsky 1995; Reynolds and Descy 1996; Bowen *et al.*, 2003; Hein *et al.*, 2003; Thoms, 2003; Thorp and Casper, 2003). It has also been well established that fish perceive patches at small (substrate) to large (e.g., floodplain areas) size ranges and use them differently depending on the season and their life history characteristics (Galat and Zweimüller, 2001; Schiemer *et al.*, 2001b). A rapidly growing area of research in rehabilitation of large lowland rivers involves studies of the impact of reduced patch complexity and decreased connectivity among lateral patches (e.g., Hein *et al.*, 2005).

Even though large patches occur in large rivers, important small patches exist within a hierarchy of larger spatiotemporal patches. These include substrate patches, which should function in the same way as in small streams. Some substrate patches develop in response to geomorphic and/or hydrologic processes, while others are formed biotically (e.g., macrophyte beds and wood snags). Macrophyte patches, which predominately occur from small open-canopy rivers to slackwaters in large rivers, can be characterized by different substrate size, oxygen conditions, benthic organic content, and nutrient relationships compared to the more homogeneous surrounding habitats of bare substrate. Their presence in large rivers is tightly bound spatially to stream hydraulics, and their patch size also fluctuates seasonally. Indeed, macrophyte patches disappear seasonally in many temperate zone rivers. Consequently, the benthic and littoral plankton assemblages associated with macrophyte beds are influenced by a spatiotemporal hierarchy of biotically and abiotically controlled patches. In some ways, small patches can become even more important in large rivers because hard substrates are often relatively more abundant in small streams than in large rivers where silt and sand often predominate. For example, the distribution of isolated rock outcroppings is vital to interactions between sympatric species of prosobranch snails in the Ohio River (Greenwood and Thorp, 2001). Water column patches of nutrients and plankton are also present in rivers, but these have not been thoroughly investigated.

Changes in the nature and variability of large and small patches often modify community composition (e.g., see references in Vannote *et al.*, 1980; Pringle *et al.*, 1988). For example, Brown *et al.* (1998) suggested that pulmonate snails are more common in temporary headwater streams and shallow littoral margins of riverine ecosystems because of a greater physiological adaptation in variable habitats, while prosobranch snails dominate the gastropod fauna of larger rivers (and in spring-fed rivers) because of greater competitive ability, lower risk of predation (from protection afforded by thick, often armored shells), and/or less-variable physical and chemical regimes.

Element I: Nested, Discontinuous Hierarchies of Patch Mosaics

Patches exist within a hierarchy of spatial and temporal scales and can be considered fundamental structural and functional units composing a nested hierarchy (Kotliar and Wiens, 1990; Wu and Loucks, 1995). Dynamic processes, including disturbances, operating at one spatiotemporal scale in a riverine ecosystem alter patterns and processes at higher and lower scales, but not necessarily to the same degree. Although the RCC was not explicitly a hierarchical model, Poff (1997) elaborated on this approach in his Categorical Niche Model where he defined scaled habitat features as *filters* that influence probabilities of occurrence by species with appropriate traits (see also Malmqvist, 2002). [In the parlance of hierarchical theory, these filters are equivalent to *holons*, or boundaries between horizontal levels (Allen and Starr, 1982; Wu, 1999).] Parsons *et al.* (2003) described the distribution of benthic macroinvertebrates within a hierarchical classification based principally on geomorphologically derived scales. Statzner and Higler (1986) also implicitly identified large-scale patches created by the physical characteristics of flow (stream hydraulics); however, they disagreed with the RCC's portrayal of streams as a steady gradient of stream hydraulics. Instead, they emphasized the ecological importance of discontinuities of flow at transitions between hydrogeomorphic patches. Within the nested hierarchy of a riverine ecosystem, the scale of temporal phenomena important to ecological processes tends to increase directly with the spatial scale, but the trajectory can be punctuated (discontinuous) rather than continuous (cf. Table 3.1).

Ecologists have identified a large range of patch sizes in riverine ecosystems, but research is needed to link spatiotemporal patches with a modeled hierarchy suitable for testing. In some cases, we also need stronger evidence that the patches we identify for motile invertebrates are really applicable for the target processes we have selected (e.g., Malmqvist, 2002). For instance, Peckarsky *et al.* (2000) demonstrated that patches seemingly appropriate for studying the dynamics of mayfly species in Rocky Mountain streams were in fact too small because processes operating at a large regional patch determined dispersal rates. This could also be an example of metapopulation dynamics at work. Moreover, Ward and Tockner (2001) suggested that hydrologic connectivity controls biodiversity at the floodplain scale; whereas at the even larger catchment scale, biogeographic factors are more important. Likewise, Grimm (1993) noted that algal biomass was strongly influenced by disturbance-related flow regime on an annual basis but by nitrogen availability during shorter temporal scales. Statistically demonstrating the hierarchical importance of different spatiotemporal patches in riverine ecosystems is not an easy task, but the problem can be approached with both experimental and modeling techniques (cf. Hughes *et al.*, 2008). Although previously it was sufficient to show that biotic interactions influenced species densities and distribution on a small spatial scale (e.g., interactions among immature insects on a rock), researchers should now be encouraged to take the next step and explicitly determine how important that process is for larger spatiotemporal patches where other independent variables operate. Ecologists should also examine how a given process varies longitudinally and laterally along the riverine ecosystem. Finally, aquatic ecologists may need to match the scale of environmental processes with the appropriate scale of biological organization from individuals to ecosystems (Thoms and Parsons, 2002).

Element II: Ecosystem Dynamics as a Composite of Intra- and Interpatch Dynamics

It has been known for at least a quarter century that the structure and the functioning of an ecosystem at any given time are the sum of dynamic, deterministic, and stochastic processes occurring within and among patches of different spatiotemporal scales (e.g., Pickett and White, 1985). This aspect of system dynamics is considered an emergent property of the ecosystem (Wu and Loucks, 1995). Emergent properties are difficult to detect if research is limited to fine spatiotemporal scales, for example, when investigating competitive interactions among periphyton scrapers on a rock without incorporating the effects of stream spates. In general, the consequences and significance of any focal-level process will not be revealed in a hierarchically organized system without also investigating the next higher (horizontal) level, nor can the mechanisms be demonstrated without studying one or more levels down from the focal process (O'Neill, 1988). This can be illustrated by examining, for example, community dynamics of riverine zooplankton. The local importance of deterministic processes can be studied with *in situ* experiments testing effects of mussel and fish predation on density and relative abundance of rotifers, cladocera, and copepods (e.g., Jack and Thorp, 2000, 2002; Thorp and Casper, 2002, 2003). Community responses that seem to involve a trophic cascade (Thorp and Casper, 2003) are evident only by studying mechanisms of predation and competition at a finer level and by understanding relative differences among major taxa of zooplankton in reproductive rates and reactions to river currents and turbulence. The reach- or FPZ-scale significance of these deterministic processes occurring within plankton patches should then be evaluated at higher hierarchical levels, such as between main channel and slackwater patches where disparate hydrodynamic forces may predominate. They could also be studied among seasons where differences in both abiotic (temperature and hydrologic regime) and biotic conditions (e.g., variable phytoplankton productivity) may alter the importance of trophic cascades. Finally, it could be crucial to understand interactions among patches that occur via corridors (an aspect that the HPD model does not emphasize explicitly), especially those that change seasonally or over longer periods in response to floods and droughts.

Deciphering codes of community regulation and ecosystem function in riverine ecosystems will probably always be a formidable challenge because these ecosystems are exceedingly complex. Riverine ecosystems feature a nested hierarchy of patches, patterns, and processes where both intra- and interpatch dynamics can play important roles. Nonetheless, researchers can often lighten the task by concentrating on adjacent hierarchical levels, where the strength and frequency of interactions with the focal level is greater than between more distant levels of the hierarchy (Wu, 1999). For example, suppose you are interested in the role of competition for space between two benthic invertebrates living on a rock in a fourth-order stream. Your initial, focal studies might involve *in situ* exclusion and inclusion experiments to identify the competitive dominant. At a lower hierarchical level, you could evaluate at the importance of species traits for this competition by asking questions such as the following: 'What are the population sizes and seasonal trajectories of the species involved?' 'Do the species have similar range of diet, resource-harvesting abilities, and habitat preferences?' And 'What constraints do they face in colonizing new substrates?' It would also be important to look at higher hierarchical levels, with questions such as the following: 'What is the effect of rock tumbling (from a stream spate) on this competition?' 'Are the species equally susceptible to predation?' 'What is the temporal pattern of seasonal droughts and hydrologic pulses, and how do they influence local competitive interactions?' It might not be necessary, however, to factor in the flow history (1–100 years) and regime (>100 years) or to understand very long-term geomorphic processes and the zoogeography and evolutionary trends of the two species, although each of those higher hierarchical levels may provide a piece of the puzzle explaining population size and distribution of both species. The resulting loss of information from limiting one's research to the focal and adjacent levels is often insignificant but will vary with processes, patterns, and scales examined. Therefore, before dismissing the importance of processes or patterns operating at much smaller or larger spatiotemporal scales or more fine or coarse hierarchical levels, scientists need to explore at least intellectually how these distant processes may impinge on their system. They must also be careful to select the correct focal level in the environmental hierarchy to address the ecological question needing an answer (Dollar *et al.*, 2007; see Figs 3.1 and 3.2).

Element III: Linked Patterns and Processes

Pattern and process are interlinked and almost always scale-dependent. By *pattern* we mean an observable configuration of living or inanimate objects or processes in space and/or time. A few of the many examples of patterns in riverine ecosystems are the alternating distribution of riffles and pools in streams, life history characteristics of species adapted for intermittent streams vs large permanent rivers, changes in functional feeding groups from headwaters to large rivers, and species replacements of microalgae colonizing rocks over time. Processes are series of actions leading to change over time and to new or recurring patterns. Some examples of processes are species evolution, riparian leaf decomposition, algal primary production, nutrient spiraling, trophic cascades, sediment transport, and flooding.

Because both patterns and processes are hierarchically nested, the selection of appropriate spatiotemporal scale has substantial effects on interpretation. Inappropriate scales can obscure patterns and the fundamental processes creating them. Indeed, it is best to examine both at several scales to look for either continuous or stepwise changes over multiple scales, as discussed in O'Neill *et al.* (1991) and Wu and Loucks (1995). If patterns vary in a discontinuous fashion across scales, this usually implies that different processes are acting to produce the pattern. For example, the species distribution pattern of a fish assemblage in a river reach among deep-water habitats, protected areas formed by wood snags, and shallow slackwaters could at one scale be in response to local hydraulic forces, riparian vegetation, and competition among species, as well as individual species adaptations. At larger spatial scales, such as the FPZ or valley, this could be affected by basin and channel geomorphic processes and hydrologic history and regime (e.g., patterns of flow and drought distribution and intensity), which alter river channel complexity and the nature and abundance of microhabitats. At broad temporal scales, evolutionary pathways and barriers to dispersal could have ultimately influenced species distribution patterns and selection of habitats in the small-scale river reach.

Pattern does not always imply process, but sometimes one can cautiously infer processes from interlinked patterns and vice versa. This can be seen, for example, in both pristine rivers (Junk *et al.*, 1989) and those with heavily modified channels (Hein *et al.*, 2003) where productivity and diversity vary with hydrologic connectivity (fluvial distance between slackwaters and the main channel or periodicity of flooding in intermittently isolated floodplain lakes). Patterns of biocomplexity across a gradient of hydrologic connectivity sometimes appear to fit an intermediate disturbance model (Ward and Tockner, 2001, based on the model of Connell, 1978). The causation of this pattern, however, may relate to numerous factors acting individually or in concert (e.g., current velocity, temperature, suspended sediments, and both predator and resource types and levels) and could vary over time. Consequently, successful inferences about causation of patterns must factor in spatial and temporal scales in hierarchically arranged ecosystems (Fisher, 1993). Extreme caution must be used, however, in extrapolating conclusions about causal mechanisms among disparate spatiotemporal scales, because regulatory factors operating at one spatiotemporal scale may be completely dissimilar from those present at another scale and because nonlinear ecological relationships and intercomponent feedbacks are common among scales (Wu, 1999).

Processes and patterns may create, modify, or eliminate one another, depending on the spatiotemporal scale examined. For example, patterns of distribution and abundance of both debris dams and beaver dams in streams influence the process of nutrient spiraling on different spatial scales by changing rates of downstream transport, but nutrient spiraling may in turn control patterns of distribution and productivity of periphyton. Likewise, the rate of primary productivity may limit density of suspension-feeding bivalve molluscs, but the distribution and abundance of exotic mussels influences the number of type of phytoplanktons (Cohen *et al.*, 1984; Caraco *et al.*, 1997).

Element IV: Dominance of Nonequilibrial and Stochastic Processes

Depending on the temporal and spatial scales of observation, ecological units within a riverine ecosystem can be interpreted as exhibiting nonequilibrial, transient, or unstable dynamics, but rarely could these open, environmentally fluctuating ecosystems be portrayed as existing in a true equilibrial state (or in earlier parlance as a stable state). However, one's view of the equilibrial status in a riverine ecosystem may depend on: (i) whether the focus is on patterns (e.g., community composition) or processes (e.g., primary and secondary production and nutrient spiraling); (ii) the level of discrimination of each; and (iii) the spatiotemporal scale examined. Furthermore, the parameter selected for patterns and processes can influence your interpretation. Indeed, the verdict could easily be swayed one way or the other if the equilibrial state of a stream community was judged by species composition (presence/absence of species), relative abundance (rank) of taxa, or absolute abundance. Neither approach is incorrect, but the result may differ with your metric of choice. As an example, a flood event in the ephemeral Sycamore Creek of the arid southeastern USA can remove more than 90% of the benthic insect biomass and much of the attached algae (Grimm and Fisher, 1989). Fisher (1993) suggested that the rapid recovery of functional processes indicates a system that is stable because of strong resilience (though weak resistance), whereas the eventual return to a similar community composition points to system that is stable because of the opposite attribute, resistance of structural biodiversity! He noted that both viewpoints were correct, depending on which observation set was selected. Other investigators focusing on the same responses but shorter timescales or the same timescale but on relative species abundances might perceive the system as clearly unstable and dominated by stochastic processes. Indeed, a reasonable hypothesis for retention of species is that the stochastic nature of hydrologic events in Sycamore Creek prevents equilibria from developing and thereby reduces chances of extirpation of local species by interspecific competition. It is relevant to note, however, that stochastic processes do not necessarily work against stability but may, instead, constitute a mechanism underlying the apparent stability of ecological systems at different scales (Wu, 1999). 'Thus, equilibrium and nonequilibrium are not absolute and context-free, but relative and scale-dependent' (Wu, 1999).

Over the last quarter century, there has been a shift in general ecological perceptions to the viewpoint that riverine ecosystems are driven chiefly by stochastic forces related primarily to floods (e.g., Ward *et al.*, 2002) or droughts (Lake, 2003; Dodds *et al.*, 2004). This pertains to both patterns (e.g., Arrington and Winemiller, 2006) and processes (e.g., Uehlinger, 2006). Stochastic processes operate across broad spatial and temporal scales in riverine ecosystems, but small-scale processes generally tend to be more stochastic and less predictable (cf. Wu, 1999). The role of deterministic processes, albeit perhaps a support or even under-study role, seems mostly restricted to smaller spatial scales where predator–prey, host–parasite, and interspecific competitive interactions operate.

Current evidence is beginning to suggest that deterministic factors are relatively more prominent in slackwater areas, whereas stochastic factors achieve their greatest importance in flowing channels where hydraulic stress is greater (Johnson *et al.*, 1995; Thorp and Casper,

2003). However, stochastic processes could also be crucial along the moving littoral border of the aquatic–terrestrial transition zone, as described in Junk *et al.* (1989). Consequently, the structure of hydrogeomorphic patches should influence the relative importance of deterministic and stochastic factors in FPZs.

A productive area for future research could entail analysis of relative importance of stochastic and deterministic processes in riverine ecosystems as one moves from cobble- to reach-sized patches for different types of FPZs. Is the importance of stochastic factors related mostly or essentially only to hydrologic patterns (e.g., flow pulse, flood pulses, and drought) or do other environmental factors play a role in streams? Are hydrologic events always or mostly stochastic, or can they act as deterministic factors at some spatiotemporal scales (cf. Sparks *et al.*, 1990; Delong *et al.*, 2001)? Does the tendency of a hydrologic event to be stochastic vs deterministic vary with stream size and/or bed motility? Do relative roles of nonequilibrial factors vary laterally and/or longitudinally by the size of the stream or hydrogeomorphic patch and the overall complexity of the FPZ? How do these relationships differ with organismal life history patterns and ecoregion?

Element V: Formation of a Quasi-Equilibrial, Metastable State

Depending on which variable one uses to evaluate equilibrium and over what spatiotemporal scales, many if not all natural rivers exist in a nonequilibrial or perhaps quasi-equilibrial state. [The latter is labeled by some as statistical equilibrium.] It is not surprising, therefore, that many ecologists question whether riverine ecosystems can ever achieve true equilibrium. This is because they are open systems subject to major hydrologic variations over several temporal scales – a characteristic that introduces substantial stochasticity within and among patches. Nonetheless, it is theoretically possible for a quasi-equilibrial, metastable state to develop at one level of an ecosystem through incorporation of nonequilibrial processes operating within multiple patches at a lower hierarchical level (Paine and Levin, 1981; O'Neil *et al.*, 1989). In one sense, this property allows naturalists to note interyear similarities in stream species richness even though relative abundances and absolute densities vary. Before proceeding further into a discussion of Element V of the HPD, it may be useful to consider definitions of two terms, metastablity and incorporation. These interrelated concepts, which form an important component of the HPD of Wu and Loucks (1995), are less familiar and probably less intuitive to many riverine ecologists.

Ecological metastability is a term derived from physics and chemistry. In physics, a metastable equilibrium can be defined as 'a state of a system that is in pseudo-equilibrium, such that a small disturbance may not disrupt the system but a larger one would render it unstable; a practical example is a ball at rest in a slight depression at the top of a hill' (Morris, 1992). From an ecological perspective, this could describe a relatively long-lived community or ecosystem that is not likely to undergo a spontaneous transformation to a substantially different state without a moderately large disturbance. This definition does *not* imply constancy of ecological conditions (such as no fluctuation in species composition), but there is a suggestion that the basic state of the system is moderately predictable at some spatiotemporal scales. A related way to interpret metastability in ecology is that nonequilibrial patch processes at a smaller scale can produce a quasi-equilibrial state (metastability) at a larger spatiotemporal scale. That is, out of chaos comes order! For instance, because of the role of stochastic forces in a stream community, it is very difficult to predict the composition of the assemblage of species on a small patch (cobble, snag, etc.). At a larger scale, such as a riffle, however, one can reasonably predict which species will occur from one time period to the next. A number of mechanisms can promote metastability, including the following five categories described in more detail in Wu and Loucks (1995): (i) spatial incorporation of localized biotic feedback instability; (ii) environmental disturbance control of biotic feedback instability; (iii) biotic compensatory control of stochastic instability; (iv) spatial incorporation of stochastic instability; and (v) absorption of ecological instability through heterogeneity.

Although metastability is in one sense a pattern, *incorporation* is a process that can produce a quasi-equilibrium. Incorporation has been occasionally examined in riverine ecology, although the term may not have been used in published manuscripts. In most cases, studies of incorporation have involved environmental disturbances as a dominant force and deterministic processes as contributing processes. This combination is well-known in terrestrial ecology from studies of gap dynamics in a forest where local fires or blowdowns can produce a shifting mosaic, steady state in the forest as a whole that allows subordinate species to coexist with the competitive dominants (e.g., Bormann and Likens, 1979; Kohyama, 2005). In riverine ecology, the principle of incorporation is often implicitly included in studies of shifting habitat or landscape mosaics (e.g., Robinson *et al.*, 2002; Lorang *et al.*, 2005; Latterell *et al.*, 2006). We can illustrate this at the riffle scale by again returning to the example of competition for space or food on a stream cobble. We can posit that species A is the competitive dominant in a riffle and should consistently drive species B to local extinction if deterministic forces rule for a sufficiently long period. The fact that species B can usually be found somewhere in the riffle, however, could be explicable if, for example, species B is: (i) more resistant to dislodgement from aperiodic stream spates; (ii) more resilient to stream spates because it is better at recolonizing denuded substrates from refuges in the hyper-rheos; and/or (iii) able to reproduce and disperse after the disturbance but before competitive exclusion occurs. Even if it is not possible to predict accurately the species composition on a particular rock because of stochastic forces, one may be able to forecast the general composition and relative abundance of species at the riffle level. Shifting habitat mosaics has been studied more recently at coarser scales in riverine habitat (e.g., Robinson *et al.*, 2002), but the general principle of incorporation still holds at broader population or small-community scales. An interesting approach would be to examine how processes of incorporation involving ephemeral habitat from wood snags (whose presence varies in time and space in response to several instream and extrinsic factors) affect the composition and stability of riverine communities.

Metapopulations

The presence of metastability at one hierarchical level does not necessarily imply its presence at another level. Hence, metastability of diverse fish communities does not mean that individual species exhibit metastability or occur in metapopulations. Indeed, ecologists and conservation biologists have just begun to explore the potential importance of species metapopulations in riverine ecosystems. Metapopulation dynamics and stability can be examined at several spatial dimensions along a four-dimensional riverine ecosystem. Research on repopulation of stream sections denuded by hydraulic spates has often emphasized recolonization from longitudinal (e.g., upstream dispersal by flying aquatic insects, swimming by amphipods and fish, and crawling by snails), lateral (e.g., interstream, cross-land migration of crayfish and adult insects), and vertical dimensions (recolonization from hyporheic refuges) (see examples in Giller and Malmqvist, 1998; Thorp and Covich, 2001). For some species, habitat recolonization in small streams constitutes intrapopulation recruitment from refugia. In other cases, however, this may represent the action of a metapopulation. The latter explanation is especially appropriate where recruitment is derived from a separate stream in the same or different riverine ecosystem or if clearly demarcated patches exist within the same stream and intrapatch recruitment chances exceed interpatch dispersal. Metapopulation models (e.g., Levins, 1969) have been applied to fish (e.g., Dunham and Rieman, 1999) and exotic mussel populations in rivers (e.g., Stoeckel *et al.*, 1997), but the theoretical suitability of specific models (e.g., the island–mainland model of Harrison *et al.*, 1988) is occasionally challenged (Gotelli and Taylor, 1999).

A more puzzling problem that potentially involves metapopulation dynamics is how plankton populations are maintained in advective environments, given the great difficulty phytoplankton, protozoa, rotifers, and microcrustaceans have in swimming against currents. Aside from likely recruitment for some taxa via resistant eggs blown from dried floodplain habitats, metapopulation studies of river plankton are beginning to focus on the role played by both small (e.g., eddies and shoreline habitats) and large patches (e.g., lateral slackwater and floodplain habitats). Hydraulic storage resulting from channel morphology may account for river phytoplankton recruitment and production, according to Reynolds and Descy (1996), and this may explain metastability in the broader riverine community. Retentivity of slackwater areas has also been shown to be vital for recruitment of 0+ aged fish, production of zooplankton (Schiemer *et al.*, 2001b), and diversity, relative abundance, and fecundity of riverine zooplankton (Thorp *et al.*, unpublished data). Similar observations led to the development of the inshore retention concept (Schiemer *et al.*, 2001a). The first step in proving that slackwater and floodplain patches generate metastability within the larger riverine ecosystem is to document a significant export of recruits into the main channel habitat. Such movement has been noted in the River Danube by Schiemer *et al.* (2001a) who found high correlations between rotifer populations in the main channel and sinuosity index values in the river (which are related to the degree of hydraulic retention). Junk *et al.* (1989) concluded that Amazonian floodplains energetically support main channel populations of fish, but the export of living and detrital organic matter from large floodplains seems insignificant in the tropical Orinoco River (e.g., Hamilton *et al.*, 1992; Lewis *et al.*, 2000) and has been questioned for the Amazon (Hedges *et al.*, 1986).

Research on metapopulations from lateral slackwaters and floodplains could be a fertile research area. Determining export rates from hydraulically complex slackwater and floodplain sites will be the first significant challenge. Another major task will be evaluating whether large patches constitute different populations and whether those in the main channel are dependent on populations in lateral patches for recruitment. This research avenue could be crucial, however, because river channelization on several continents has reduced lateral patch complexity and severely impinged on ecological integrity in a high percentage of lowland river ecosystems (e.g., Hein *et al.*, 2005). Finally, stream ecologists need to determine which models of population sustainability are most appropriate (e.g., see Gotelli and Taylor, 1999).

THE RES AS A RESEARCH FRAMEWORK AND FIELD APPLICATIONS OF HIERARCHICAL PATCH DYNAMICS

In addition to components involving FPZs and model tenets, the RES provides a *research framework*, or tool for aiding riverine research. It combines HPD with river science (including landscape ecology, lotic ecology, and fluvial geomorphology; see Fig. 1.2). This framework is equally applicable to fundamental and applied research, although some aspects will be of greater applicability to one focus or the other. Three chapters of this book discuss how to delineate hydrogeomorphic patches (Chapter 5), the applicability of model tenets/hypotheses in altered rivers (Chapter 7), and practical applications of the RES in management and conservation settings (Chapter 8). To supplement these chapters, we briefly discuss below how to apply the seemingly esoteric HPD model to field situations. The discussion is based on both general principles of hierarchical analysis and some specific aspects of the HPD model. We believe the HPD concept is sufficiently applicable to a wide range of research projects that we *strongly* encourage our own graduate students to read the more detailed article by Wu and Loucks (1995).

River scientists are unlikely to argue against the premise that both hierarchy theory and patch dynamics models are applicable to research studies, rehabilitation, and management of riverine ecosystems, at least in some measure. Why then are these two concepts rarely incorporated into riverine studies? The answers probably relate both to research funding/publication pressures and to either ignorance (not stupidity) or intellectual laziness. (Note, the authors of this book make no claim to being totally immune to all these problems.) Complex research questions at multiple hierarchical levels demand higher research budgets and longer study periods. The most practical way of partially circumventing this dilemma – but an approach that few scientists can or will adopt – is to conduct a series of overlapping, separately funded projects over multiple years or even decades and then synthesize the results into a comprehensive model. This is a very challenging task in our current scientific atmosphere where researchers and our funding agencies tend to jump from one newly popular research bandwagon to another without really solving the original question.

By not incorporating HPD into our research, however, we become susceptible to the computer adage of 'garbage in, garbage out!' This can be illustrated with examples related to conclusions on the relative importance of stochastic and deterministic processes in rivers. Suppose we invest substantial time and money to devise elaborate descriptive and experimental studies of competitively linked, habitat distribution of fish in a single reach. If we ignore more stochastically driven processes operating at greater temporal scales while conducting this empirical research, we risk missing the effects of either floods, which can redistribute habitat patches, or droughts, which can alter longitudinal connectivity (e.g., by changing a continuous channel into isolated pools). In either case, competitive advantages and pressures could be altered dramatically within the fish community. Does this mean that studies of deterministic processes and finer spatiotemporal scales are unimportant? Certainly not! For example, some species have adapted to fluctuating environments by exploiting refuges in space (thereby avoiding high current velocities or loss of water) or time (e.g., periods of reproduction and competitive interactions that are much shorter or longer than the likely period of environmental fluctuations). In these spatiotemporal contexts, deterministic processes could prove pivotal. As an example cited earlier, stochastic and relative constant hydrologic and hydraulic processes may dominate zooplankton population and community dynamics in the main channel of large rivers, while deterministic processes may be instrumental in slackwaters (Thorp and Casper, 2003).

Hierarchical patch dynamics can be employed in numerous ways to improve both fundamental ecological research and applied management and rehabilitation of riverine ecosystems. The most widely applicable component involves hierarchical analyses at multiple spatial and temporal scales, with an emphasis on the levels immediately above and below the focal question level. It may initially seem that the HPD model is intrinsically more useful in fundamental than applied projects, but this may partially reflect its historically more common and overt use in basic research. In developing policies on artificial flow regulation, for example, the initial approach by governments was to regulate the minimum flow based on a constant discharge value. Following the publication of the natural flow regime paradigm (Poff, 1997), it has become widely acknowledged that variability in flow is a critical process in riverine ecosystems. This has led to a growing trend, for example, of intraseasonal, fluctuating patterns of water discharge from reservoirs for the sole purpose of improving the ecological health of the downstream aquatic environment (Tharme, 2003). A temporal hierarchy of release patterns where stochastic (acyclical) flow patterns replace cyclical release (i.e., consistent pattern of variable flow) has not yet been widely incorporated into management practices on either seasonal or multiyear scales, but such patterns could probably improve environmental responses (Thoms and Sheldon, 2000). This reflects the conclusion of Element IV of the HPD model that nonequilibrial conditions and stochastic processes play a dominant role in ecosystem stability (or metastability, as described in Element V). On a large spatial scale, we need to understand how to manage different portions of a riverine ecosystem to achieve preferred biotic structural (e.g., enhanced species richness) and functional responses (e.g., enhanced nutrient processing to reduce estuarine export of nitrogen). This will involve understanding how regulatory actions by governments influence complex relationships between patterns and processes at different hierarchical levels (a component of HPD Element III).

Integrating the theory of patch dynamics into research can also improve fundamental and applied research, though perhaps to a lesser extent than for hierarchy theory. Ecologists need to avoid focusing their efforts at a single patch and should instead incorporate the variability in space and time among patches, especially when working much below the spatial level of FPZs. This dynamic process becomes crucial for conserving species that exist within metapopulations. Conservation of species at large spatial levels may also require a focus on a regional riverine landscape or an entire riverine ecosystem composed of large hydrogeomorphic patches. Dynamic hydrogeomorphic patches composed of aquatic and terrestrial elements (and those that fluctuate between the two states) are essential elements of riverine landscapes, but at present we lack a fundamental understanding of how they fit together ecologically in a variable environment. Consequently, we have difficulty in shaping management practices affecting spatial extent and temporal variability of these riverine landscape components. Questions of how far to set back levees and how often to reconnect wetlands should both be influenced by the hydrogeomorphic nature of the large FPZ patches.

From a generic perspective, the best advice we can give riverine scientists at all training levels is to include at least four attributes of the HPD model in their project design (not necessarily in order of importance). First, recognize that the parallel hierarchies of geomorphology, hydrology, and freshwater ecology have different organizational elements and levels. Once this is accomplished, spatial and temporal scales for each level of organization for the different discipline hierarchies should be assigned. This allows individual parts to be distinguished by different frequencies of occurrence and/or rates of change. Second, select the appropriate spatiotemporal scale of the environment to match the focal organizational level of the ecological question (see Fig. 3.1). Third, work with multiple patches and examine dynamic relationships within and among patches, and ask how your results might be affected if the project design were at a different spatial or temporal scale of the environment (would that affect the processes or patterns you need to study?). Finally, consider how your results could be altered if the relative roles of deterministic vs stochastic processes changed based on factors listed above. All these steps should promote the development of alternative hypotheses and lead to a better understanding of the environment.

The Spatial Arrangement of River Systems: The Emergence of Hydrogeomorphic Patches

INTRODUCTION

Rivers are complex landscapes. Their physical structure (morphology), the spatial arrangement of components that make up riverine landscapes, and their behavior over time are the result of a host of interacting influences, operating at multiple scales. Influencing factors can be grouped into independent catchment and independent channel variables (Morisawa, 1968). Climate, geology, soil character, vegetation, and land use are independent catchment variables that determine the flow and sediment regimes of river systems, in particular the volume, timing, and character of water and sediment supplied from the catchment. The flow and sediment regimes are independent channel variables that primarily govern the size, shape, slope, and planform configuration of the river channel and adjacent floodplain - all dependent channel variables. Interactions between independent catchment and channel variables and their influence on dependent channel variables are depicted in Fig. 4.1. Although other factors such as local climate and the nature of riparian vegetation may also influence the physical structure of river channels (e.g., Naiman *et al.*, 2005), these independent variables and their interactions shown in Fig. 4.1 provide the main framework for riverine landscapes. The nature and degree of influence of independent factors also changes along a river. Some influences may be subtle, such as sediment transformations in a long, slowly aggrading river channels, while others are abrupt, such as at major tributary junctions. Hence, a diverse array of physical structures may exist between and within riverine landscapes.

Broad-scale patterns in the spatial arrangement of riverine landscapes have been observed despite variations in the influence of independent catchment and channel variables. The early work of Schumm (1977a) described a general downstream transition within rivers from a production zone in the headwater regions through a transfer zone to an accumulation zone along lower river reaches. This clinal or continuum perspective has been embraced and extended by many (e.g., Palmer, 1976) by incorporating scalar perspectives on patterns in four dimensions (longitudinal, lateral, vertical, and temporal – sensu Ward, 1989). For example, the ecogeomorphological approach proposed by Davies *et al.* (2000) explained the continua of

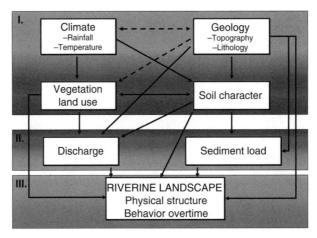

FIGURE 4.1   Factors influencing the physical structure and behavior of riverine landscapes over time (modified from Morisawa and Laflure, 1979). Three levels of influence are recognized: (I) independent catchment factors; (II) independent channel variables; and (III) dependent channel variables as described by Schumm (1997a). (See color plate 3).

physical structures and their behavior in time and space at different scales along the lower Zambezi River. This approach focused on longitudinal patterns and included lateral and vertical patterns to illustrate complex responses of the entire river system to flow regulation. Biological implications of physical gradients in river systems, as noted in earlier chapters, have also been articulated by many, the most notable being that developed in the RCC (Vannote *et al.*, 1980).

Deviations from broad-scale clinal patterns in the spatial arrangement within river systems have been reported (cf. Montgomery, 1999). Local controls can have significant influences on the composition and spatial arrangement of physical structures in rivers, and it has been suggested that these can be more important than the position within a river network (e.g., Junk *et al.*, 1989; Thoms *et al.*, 2004). Although acknowledging the importance of flow and sediment regimes on the composition and spatial arrangement of physical structures within river networks, additional themes have emerged portraying river systems as a mosaic of hydrogeo-morphic patches (cf., Thoms, 2003; Thoms *et al.*, 2004; Thorp *et al.*, 2006). Applying an HPD approach (*sensu* Wu and Loucks, 1995) to the organization of river systems challenges many of the traditional views, concepts, and models of riverine landscapes. This approach recognizes and embraces the dynamic nature of riverine landscapes at multiple scales rather than accepting a systematic organization of their character. Rivers are continually evolving in response to longer-term natural disturbances, shorter-term pulsed or ramped human interventions, and high-magnitude, low-frequency episodic events (Lake, 2000). Responses may be cyclic, nonlinear, and/or lagged (Phillips, 2006), and the effects of single or multiple disturbance(s) may overlap and interact, thereby increasing the probability of alternative states in river systems (Reid and Ogden, 2006). This view of riverine landscapes moves away from notions of equilibrium and cyclic behavior as a means to explain nonlinear relationships and stochasticity. Recent views of river systems (Dollar *et al.*, 2007) emphasize the importance of scale and their hierarchical organization, acknowledging both top-down constraints and the emergent features of bottom-up influences. Recognition of interacting processes is a central tenet of hierarchy theory – something that has been long recognized in geomorphology (*sensu* Schumm and Lichty, 1965) but not totally integrated into the epistemology of river systems thinking. Contrasting views of the clinal and patch approach to the spatial arrangement and behavior of riverine landscapes reflect the differences between Darwinian and Newtonian approaches to science in some ways (Harte, 2002). The former embraces the principles of complexity, contingency, and interdependence, while the latter strives for simplification, ideal systems, and predictive understanding.

In this chapter, concepts and theoretical approaches to the spatial arrangement of riverine landscapes are discussed, and this is followed by a review of the river characterization. In particular, the results of a meta-analysis of the more commonly used river characterization schemes are presented. A characterization scheme for riverine landscapes then presented, and its application demonstrated with two case studies: the Murray–Darling Basin, Australia and some of the rivers in the Kingdom of Lesotho, Africa.

THE SPATIAL ARRANGEMENT OF RIVERINE LANDSCAPES

Approaches to studying the spatial arrangement of river systems are many and varied. Most have encompassed the notion of gradients in different dimensions, while others incorporate *interalia*, discontinuities arising from local scale factors as well as functions at multiple scales. Longitudinal patterns in the physical character of river channels at the catchment scale have been recognized for decades and have formed the theoretical basis in much of fluvial geomorphology. These patterns are usually explained in terms of progressive changes in the flow and sediment regime of rivers, commonly in the downstream direction. As a consequence, rivers were initially partitioned into zones of production, transport, and accumulation and were illustrated by being characterized by different geomorphological attributes and behavior (Schumm, 1977a). In idealized river channel networks, stream power (a measure of the ability of a river system to undertake work), which is usually measured in terms of sediment load, has been shown to change in a predictable manner downstream (Church, 2002). Physical gradients within riverine landscapes not only exist in the longitudinal dimension but also are thought to extend to the lateral, vertical, and temporal dimensions (Ward, 1989). These gradients result from directional fluxes and interactions between different components of river systems.

Fractal relationships are common to natural systems (Dodds and Rothman, 2000), and allometric relationships describe how the dimensions of different parts of a river network grow with respect to each other (Mandelbrot, 1982). In river systems, the research of Hack (1957) and Horton (1945) represents early studies in this area, and the laws derived from their work relate stream order, the number of streams (hence stream density), and stream length to upstream catchment area and hillslope gradient. The exponents of the ratios derived by Hack and Horton differ between catchments, thereby describing the shape and degree of landscape dissection of different river systems. Recently, Rodriguez-Iturbe and Rinaldo (1997) demonstrated that river systems can self-organize in optimal patterns resulting from the propensity of water and sediment to maximize the energy dissipated in transporting sediment and water to an outlet. However, natural river systems are suboptimal, with local factors disrupting the idealized patterns. Thus, fractal relationships in network geometry can fluctuate quite widely within a given system and can vary depending on the scale of resolution by the observer (Dodds and Rothman, 2000). Critics of clinal and fractal models of river systems point out that in many systems, the expected patterns cannot actually be observed at some or all scales (Callum *et al.*, in press).

River systems can break theoretical rules in terms of their spatial arrangement and behavior over time. Often patterns of physical character contradict concepts of environmental gradients and network geometries, thus defying attempts to predict their spatial arrangement. Studies have shown that rivers display abrupt changes in hydraulic character, morphology, and biology, rather than exhibiting gradual changes as supported by the notion of a continuum (Townsend, 1989; Montgomery, 1999; Poole, 2002). The RCC (Vannote *et al.*, 1980) assumed that the spatial arrangement of physical characteristics within a river network forms a continuum, but new perspectives on spatial arrangements emphasize discontinuity and patchiness. The serial discontinuity concept of Ward and Stanford (1983b) addressed the effects of dams, which reset the longitudinal continuum, thereby highlighting abrupt transitions between river segments with dissimilar physical structure. Likewise the FPC (Junk *et al.*, 1989) suggested that in large flood-plain settings, river systems are shaped more by lateral exchanges of water, sediment, and nutrients rather than by upstream processes. Hence, the character and the behavior of large floodplain rivers do not always conform to a longitudinal continuum in time or space (Thoms *et al.*, 2005). Discontinuities are often associated with regional or local variations in climate, geology, riparian conditions, tributaries, lithology, and/or geomorphology or with human inter-ventions that disrupt the flow, sediment, and/or disturbance regimes. Discontinuities and devia-tions reflect breaks in network connectivity, whereas strong clinal patterns are dependent on strong connections between elements within the riverine landscape (Callum *et al.*, in press).

Clinal perspectives generally assume pristine conditions where a dynamic equilibrium state can be attained. However, concepts of stability and equilibrium are scale-dependent (Turner *et al.*, 1993). Patterns in the physical and biological character observed within riverine land-scapes differ at different scales (Parsons and Thoms, 2007). Different processes control the composition and configuration of elements within a riverine landscape at different scales. In addition, human influences have also significantly altered natural processes so that stochasticity factors become involved. As a result, outcomes of river behavior cannot be predicted with certainty. Indeed, rivers are better conceptualized as complex adaptive systems (Dollar *et al.*, 2007) characterized by multiple interactions, feedbacks, and nonlinear relationships, with responses being induced by exceeding thresholds (Levin, 1992). Changes within riverine land-scapes are path-dependent, being strongly influenced by the history of a specific location. Therefore, deterministic models cannot fully describe the entire range of variability within riverine landscapes. In order to deal with scale, complexity, and variability, many have adopted the paradigm of HPD when describing or explaining spatial and temporal relationships (Wu and Loucks, 1995).

The complexity of riverine ecosystems challenges many traditional scientific methods. Their multicausal, multiple-scale character limits the usefulness of the conventional reductionist-falsification approach, except when applied at very small scales and within limited domains. Hierarchy theory is used widely in ecology and earth sciences, providing a paradigm that addresses the complexity of natural systems and allows the integration of knowledge from diverse disciplines (Allen and Starr, 1982; O'Neill *et al.*, 1986). It is a useful approach for interpreting river complexity. A hierarchy is a graded organizational structure. A particular hierarchical level (or holon) in a system is a discrete unit of the level above it and an agglomera-tion of discrete units of the level below it. A particular level in the hierarchy exerts some constraint on lower levels (O'Neill *et al.*, 1986), especially the one immediately below; lower levels, conversely, can influence the structure and functioning of higher levels. In consideration of a particular level, the downward constraints and upward influences of all other levels are encapsulated by the characteristics and properties of the levels immediately above and below only. The simultaneous operation of processes at different levels, within particular contextual constraints, gives rise to emergent properties.

Applying hierarchy theory to natural ecosystems, Wu and Loucks (1995) developed the paradigm of HPD. In essence, this model recognized that the temporal and spatial heterogeneity of ecosystems can be described in terms of observable mosaics generated by different dominant processes in different areas of a landscape at different scales. The disparate patches that compose the mosaic can be considered as a process domain, characterized by dominant pro-cesses that interact to produce the observed pattern (Montgomery, 1999). Homogenous patches occur when the nature and rate of the dominant process(es) is similar across the whole patch. Comparable patterns observed across several scales (i.e., a fractal pattern) implies the domi-nance of these processes across scales, indicating limits at which cross-scale extrapolations are valid (Callum *et al.*, in press).

Hierarchical patch dynamics also suggests that the distribution of dominant landscape processes can be revealed through an analysis of spatial and temporal landscape patterns and connectivities (Wu and Loucks, 1995). However, the link between an observed pattern and its presumed causal process regime is often difficult to demonstrate because of feedback processes at multiple scales (Parsons and Thoms, 2007). Thus different initial conditions and histories may result in different patterns in areas that experience similar process regimes. Conversely, similar patterns may result from several different process regimes (e.g., Schumm, 1991). In addition, different processes may operate on the same patch at different times.

Hierarchical patch dynamics thus offers a paradigm in which the higher levels of organization impose structural and functional constraints on lower levels, while local heterogeneity can emerge from finer-scale interactions within and between patches. Although the continuum perspective tends to focus on landscape-scale connectivities, HPD invites consideration of linkages between system elements at different scales. At scales characteristic of a hierarchical level of organization, linkages and interactions between system elements are most numerous and intense (Allen and Starr, 1982). Cross-scalar links are also seen, notably in the ways in which lower levels are contextually constrained by higher levels.

RIVER CHARACTERIZATION

Characterization requires the ordering of sets of observations into meaningful groups based on their similarities or differences and then attaching labels to these groups. Implicit in this exercise is the assumption that relatively distinct boundaries exist, and these may be identified by a discrete set of variables. Although river systems are continuously evolving and often display complexity, the grouping of a set of elements with a definable structure can aid in relating broader physical factors. It also assists in understanding why riverine ecosystems have certain biological characteristics. There are many geomorphic river characterization schemes, reflecting the diversity of environmental and contextual settings and the variety of potential approaches to ordering complex systems. Motivations for identifying different types of rivers are varied, but Kondolf *et al.* (2003) suggested that there are two main objectives for river characterization, which are:

1. To improve our scientific understanding of how rivers function and whether they are clustered in homogeneous classes; and
2. As a basis to guide management and informed decision making.

Therefore, the characteristics of rivers emphasized in various approaches vary widely and depend in part on the purpose of the characterizer. Mosley (1981) outlined eight fundamentals for any characterization exercise but noted that there are few schemes or approaches that conform to all these principles.

- Characterization should be designed for a specific purpose.
- Objects that differ may not easily fit into the same characterization.
- Characterizations are not absolute but will change with more or improved knowledge.
- Differentiating factors should be properties of the objects being characterized rather than factors assumed to affect or determine the objects.
- Characterization should be exhaustive and the classes should be exclusive.
- Characterization should proceed at every stage of characterization as far as possible on a single characteristic.
- Differentiating factors must be important or relevant to the purpose of characterization.
- Factors used to characterize at the broader scales must be more important for the purpose of characterization than those at finer scales.

Fluvial geomorphology provides an ideal starting point with which to assess the physical structure and function of riverine landscapes. Without a geomorphological understanding of the distribution of physical processes within a river (processes that determine river structure), approaches employed to understand and determine how rivers function are unlikely to succeed in a sustainable manner over the long term. Rivers are complex systems that demonstrate diverse character, behavior, and evolutionary traits both between and within systems. Given this variability, individual riverine ecosystems need to be approached and managed in a flexible manner on the basis of what is actually present and happening within that river.

In a general review of the geomorphic classification of river systems, Kondolf *et al.* (2003) recognized five different broad characterization groups. The first group is simply termed early characterization schemes. These are mainly based on their genetic relation to geological structure and/or the evolution of landscapes. The geographic cycle of Davis (1899) is a good example of the latter, as it was derived from evolutionary theory in which rivers were divided into three stages of an evolutionary cycle – mature, youthful, and young. Process-based characterizations form the second group, and there are numerous examples of these. The initial work of Leopold and Wolman (1957) presented a quantitative basis for differentiating braided, meandering, and straight channels based on relationships between slope and discharge. Indeed, this pattern-based approach formed the basis for many other river characterization schemes including those by Schumm (1977b), Nanson and Knighton (1996), and Church (1992), to name just a few. The third group is based on stream power approaches. These recognize that stream power is a key variable in shaping river channels, and commonly focus on discontinuities in specific stream power as a means to delineate between different streams and reaches. However, imprecision in the calculation of stream power and the lack of clear thresholds between stream patterns of different stream power question the utility of using this factor as the sole variable of differentiation. However, many approaches use stream power in combination with other factors for river characterization purposes. Characterizations have been devised explicitly for river management, the fourth group of river characterization approaches. These approaches are generally single-scaled and are focused on the specific needs of the management context. The fifth group of characterization approaches is composed of those employing hierarchical methods, and there are many (e.g., Frissell *et al.*, 1986; van Niekerk *et al.*, 1995; Montgomery and Buffington, 1998; Thoms *et al.*, 2004). This group of approaches recognizes that broad-scale parameters determine the range of behavior of physical processes at smaller scales. There is an interaction between physical units of a river system at different scales, and this determines the character and behavior of river systems (Naiman *et al.*, 1992). Using a hierarchical or a scalar approach to the characterization a river system can be examined across a range of scales, with each scale having a different degree of sensitivity and recovery time to disturbances such as floods or droughts. Such an approach not only defines the structural components of a river but also recognizes the relative importance of factors controlling the long- and short-term behavior of river (habitat) changes within a spatial scale.

Very few quantitative analyses of river characterization schemes are available. However, recent reviews by Boys *et al.* (2005) and Rayburg *et al.* (2006) highlighted the utility and limitations of various schemes. Characterization schemes can be seductive, especially for the nongeomorphologist. They often simplify geomorphological analyses to assist the issue at hand (Kondolf *et al.*, 2003), and many fail to employ a full hierarchical arrangement framework. Characterization approaches must be cognizant of scale in general and the purpose for which the characterization scheme will be employed. To evaluate the nature of characterization schemes, the metadata from Boys *et al.* (2005) and Rayburg *et al.* (2006) are combined in our analysis below. We use 152 river and floodplain characterization schemes and reevaluate them, each with different scales of operation, approaches, data utilizations, and applicabilities. These schemes include the most common and successful river characterization endeavors reported in the international literature. Each scheme was deconstructed to look at its basic properties with respect to:

- the common spatial scale or scales employed;
- the incorporated discipline(s);
- various methods used to collect the characterization data;
- the general approach (e.g., qualitative or quantitative, form- or process-based); and
- the primary focus within the riverine landscape (e.g., river channels, floodplains, or channel networks).

Five spatial scales were used in the meta-analysis: basin, river system, reach, subreach, and site scales. Indicative data types or variables that might represent each spatial scale were recorded, and these included geology, soils, climate at the basin scale, valley width or confinement, and slope at the river system scale. Variables at the reach scale included cross-sectional morphology, riparian vegetation, planform, discharge, and bedforms (e.g., pools and riffles, bars, islands, and step pools) at the subreach scale. For the site scale, riverbed sediment texture, water chemistry, and hydraulics were noted. For each characterization method, a mark was given for the scale at which data were collected. This allowed any given approach to be uni- or multiscaled and to incorporate any combination of scales, up to and including all five. The discipline that an approach used to characterize rivers was assessed according to the data types adopted in the method. Five discipline areas were identified. These were the following: (1) hydrology, which commonly used discharge or flow data; (2) hydraulics, which included flow characteristics such as velocity, shear stress, and stream power; (3) geomorphology, which used river channel cross-section morphology, river channel planform, and bed sediment size; (4) aquatic ecology, which included fish, benthic macroinvertebrates, zooplankton, and diatoms; and (5) chemistry, which commonly used basic water data information and/or other chemical variables. Three possible data collection methods were identified, which are the following: (1) field data, which require the collection from an actual site; (2) desktop data, which can be obtained from existing sources, such as flow gauge data, climate data, geology, and/or a combination of topographical maps and digital elevation models (DEMs); and (3) remotely sensed data derived (or can be derived) from remotely sensed imagery (e.g., aerial photography and satellite imagery). As with the spatial scale and discipline criteria, characterization techniques may use multiple collection methods to compile required data.

The approach of each characterization technique was incorporated in two ways. First, the manner in which the groups were defined was considered; and second, three categories were developed to represent this, which are as follows: (1) quantitative approaches, where classes (or groups) are defined using numerical or statistical techniques; (2) semiquantitative approaches, where the classes (or groups) have numerical boundaries but these are determined a priori, not mathematically or statistically; and finally (3) qualitative approaches that rely on subjective or descriptive methods for determining classes (or groups), i.e., numerical methods are absent. The second means for incorporating the approach of a characterization scheme was to consider its focus on form and/or process. This is described using four groups, which are as follows: (1) form, where the classes are based on variables such as river channel width, riverbed sediment size; (2) process, where classes are based on process variables (e.g., erodibility and energy); (3) form to process variables, which are primarily form based but are used to infer process; and, (4) process to form variables, where classes are process-based but are used to infer form. Combined, these subcategories define the approach used within a characterization system. Unlike the previous descriptive criteria, these are exclusive categories. In other words, a technique may be quantitative, semiquantitative, or qualitative but not more than one of these.

The last of the descriptive criteria employed in this meta-analysis of characterization schemes is the focus of the method itself. Three categories were recognized to describe the focus, which are as follows: (1) channel, in which the technique incorporates channel variables; (2) floodplain, where the technique incorporates floodplain variables; and (3) channel network, where the technique considers stream networks. Here any characterization scheme may include one or more foci.

The meta-analysis of the 152 classification schemes assessed according to the five key criteria is provided in Table 4.1, and quantitative analyses of the results are presented in Fig. 4.2. The majority of river characterizations (75%) are undertaken at multiple scales, and the most common scales employed are the reach and site scales (see Fig. 4.2a). Characterizations also employed data from the remaining three spatial scales but to a lesser extent. Further analysis of these data also reveals that most river characterization schemes (83%) used information at only two spatial scales – most commonly the reach and site scales. It is also pertinent to note that very few of the schemes reviewed (<3%) used information collected at more than three scales. In addition, 95% of all schemes considered had a predominantly geomorphological basis of accounting. The remaining four disciplines were approximately equally represented at about 20% (see Fig. 4.2b). Single- and multi-discipline approaches were about equal with half being single discipline (most commonly geomorphology) and half containing data from two or more disciplines. Methods for collecting data and information used in various

TABLE 4.1   Criteria Used to Evaluate Different River Classification Schemes

Criteria	
Spatial scale	• Basin: geology, soils, climate • River system: valley width or confinement, slope • Reach: cross-sectional morphology, riparian vegetation, planform, discharge • Subreach: bedforms (e.g., pools and riffles, bars, islands, step pools) • Site: bed sediment size, macroinvertebrate assemblages, water chemistry, flow hydraulics
Discipline	• Hydrology: discharge • Hydraulics: flow characteristics including velocity, shear stress, and stream power • Geomorphology: shape characteristics including cross-sectional shape, planform, and bed sediment size • Ecology: aquatic ecology including fish and macroinvertebrates • Chemistry: water quality or other chemical characteristics (e.g., dissolved oxygen, pH)
Data collection method	• Field: the data require a field visit to collect (e.g., chemistry, biota) • Desktop: data obtained from existing sources (e.g., flow gauge data, climate data, geologic and/or topographic maps, DEMs) • Remote sensing: data derived from remotely sensed imagery (e.g., aerial photography, satellite imagery)
Approach	• Quantitative: the classes (or groups) are defined using numerical or statistical techniques • Semiquantitative: the classes (or groups) have numerical boundaries but these are determined a priori, not mathematically • Qualitative: the classes (or groups) are subjective (or descriptive) with no numerical boundaries The second means for incorporating the approach of a classification scheme was to consider its focus on form and/or process • Form: the classes are based on form variables (e.g., width, sediment size) • Process: the classes are based on process variables (e.g., erodibility, energy) • Form to process: the classes are form-based but they are used to infer process • Process to form: the classes are process-based but they are used to infer form
Focus	• Channel: the technique incorporates channel variables • Floodplain: the technique incorporates floodplain variables • Network: the technique considers stream networks

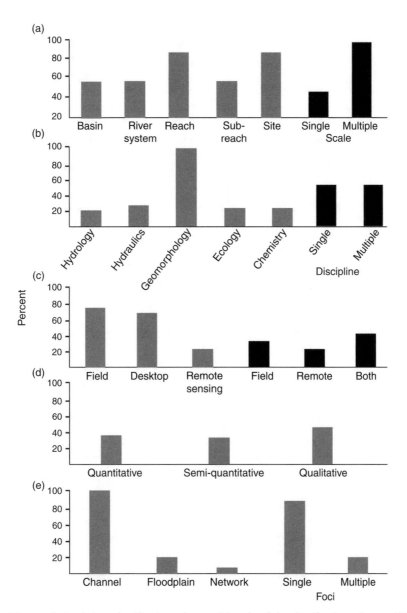

FIGURE 4.2 Meta-analysis of river classification schemes: (a) scale of the classification schemes; (b) discipline of various classification schemes; (c) type of data used; (d) nature of data analysis; and (e) physical focus of the classification schemes.

schemes are presented in Fig. 4.2c. The majority relied on data and information collected from the field (76%) compared to desktop-based data collection methods (67%) or those that relied on remotely sensing data (21%). The most common overall approach was for a combined data collection method using field and desktop or remotely sensed data and information (44%) with only 32% of techniques using field components alone and 24% of techniques using only remote (desktop or remote sensing) data collection methods.

A summary of the approaches adopted by the different characterization schemes is presented in Fig. 4.2d. Most approaches were strictly qualitative (42%) relying on descriptions, general patterns observable by eye, and the experience of the characterizer. Semiquantitative approaches accounted for 30%, while only 27% of the approaches had a purely quantitative approach. Thus, the minority of the methods reviewed were without some subjective component. Furthermore, the proportion of characterization techniques categorized as semiquantitative is inflated because a method need to have only one variable with a quantitative boundary to be labeled as semiquantitative. Qualitative approaches often make it difficult for nonexperts to characterize rivers because they require some level of expert interpretation to help place the rivers in different categories. Finally, Fig. 2e presents the focus of each of the 152 characterization techniques reviewed. In excess of 90% of the techniques had a channel focus, while only 18% included the floodplain and 6% focused on the entire channel network. Moreover, the majority of the techniques (83%) focused on only a single aspect of river character (usually channel), while those with multiple foci typically included both channel and floodplain components.

From this meta-analysis, it appears that very few of the commonly used schemes for river characterization employ a hierarchical approach. Furthermore, some of those espousing such an approach do not apply hierarchy theory correctly. As previously noted, a hierarchy is a graded organizational structure, and the character of any level within this system is a discrete unit of the level above it and an agglomeration of discrete units of the level below it. Thus, those schemes using information from vastly different levels or scales and not levels or scales immediately adjacent to one another within a hierarchical organized system are incorrectly using the application of hierarchy theory for river characterization.

There are, however, approaches based on sound conceptual models of river systems that recognize the array of spatial arrangements within river systems at multiple scales. The fluvial hydrosystem concept of Petts and Amoros (1996) was one of the first to attempt this and formed the basis for others such as the river framework approach of Dollar *et al.* (2007). Both seek to describe patterns in riverine landscape in four dimensions (Ward, 1989) and at different scales. The spatial arrangement of both physical and biological elements is considered to be largely determined by the flow and sediment (both organic and inorganic) regimes. Functional and genetic links between adjoining components of the riverine landscape result in clinal patterns conceptualized as continua by allowing cascading interactions and the transfer of matter and energy between landscape units. However, the integrity of river systems depends on the dynamic interactions of hydrological, geomorphological, and biological processes acting in longitudinal, lateral, and vertical dimensions over a range of timescales. River systems also relate to the variability of the hydrological, geomorphological, and biological processes that determine both the types of physical patches present and the magnitude, duration, and frequency of their connectivity. Thus, resultant interactions may also produce riverine landscape mosaics rather than a system characterized by gradients.

A CHARACTERIZATION SCHEME FOR THE RES

Using a fluvial hydrosystem concept, Petts and Amoros (1996) proposed a five-level approach for the characterization of riverine ecosystems. This has been further developed by Thoms *et al.* (2004) and Dollar *et al.* (2007) into a seven-tier organizational hierarchical framework for river systems, as shown in Fig. 4.3. The fluvial hydrosystem considers the catchment to be the primary unit for investigating river systems. Its structure is a result of geological processes and climatic influences operating at times scales over 10^4 years. Nested within catchments are riverine landscapes located within the topographic lows or the catchment valleys. Riverine landscapes are composed of two components: the riverscape and the floodscape. The riverscape contains the active or bankfull channel, the riparian zone, and those marginal areas of the active channel that interact frequently with it and may include features such as side channels. The floodscape is that area of the landscape that has been formed as a result of alluvial processes and contains both an active and an inactive floodplain, as defined in

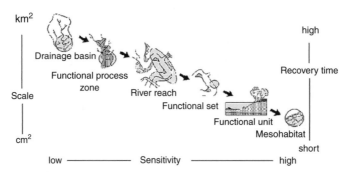

FIGURE 4.3 The hierarchical organization of riverine landscapes.

terms of the nature of its inundation by the adjacent river channel. The latter can contain terraces or abandoned floodplains that are no longer inundated by the contemporary river system. Thus, the floodscape will have an array of predominantly terrestrial to aquatic ecosystem processes as well as flora and fauna associated with it. Functional process zones (FPZs) are lengths of the river system that have similar geological histories, discharge, and sediment regimes. Two common examples of FPZs are straight-channel zones in confined valleys and low–gradient, highly sinuous, meandering channels with broad floodplains. Channel pattern is an obvious difference between FPZs; however, they also have contrasting flow and sediment regimes. The scale of FPZs reflects the size of the river system being studied and can vary from several kilometers to over 500 km (cf., Thoms et al., 2004). River reaches are repeatable lengths of river channel within a process zone and, therefore, have a similar river channel style. They are typically based on river channel planform or bedform character, with a reach being delineated as a number of meander bends or riffle–pool sequences. In many biological studies, for example, a reach is defined as a riffle–pool sequence, while many fisheries ecologists delineate a river reach as several meander bends. Each river reach may be divided into functional sets of typical units associated with specific landforms within the riverscape or the floodscape. Typical landforms may include side channels or anabranches, wetlands located within the floodscape, cutoffs, and the main active channel. The character of each functional set within the riverine landscape is determined by the magnitude, frequency, and duration of longitudinal, lateral, and vertical fluxes, which are related to the geomorphology of each landform. Petts and Amoros (1996) suggested that the appropriate timescale for analyzing functional sets is between 10 and 1000 years. A functional unit is indicative of the physical conditions at a smaller scale. They commonly represent physical habitats associated with animal and plant communities present at a site and may range in size from 100 to 10,000 m^2. Essentially all units within a functional set evolve from a single origin by progressive changes over different time periods (Petts and Amoros, 1996). Functional units can be subdivided into mesohabitats. These are sensitive to variations in flow, sediment, and nutrient fluxes and as a result may change yearly. Common mesohabitats include sand and gravel bars, in-channel benches, scour holes, gravel patches, undercut river banks, and other smaller features such as emergent and submerged vegetation, submerged wood, and other substrates.

APPLICATION OF THE CHARACTERIZATION FRAMEWORK

Using this framework, rivers can be characterized through an analysis of a series of data such as topographic maps, satellite and airborne imagery, geological surveys, sediment yield, and discharge data. Chapter 5 provides more detail for undertaking a river characterization exercise.

Two examples are presented, the Murray–Darling Basin in Australia and several rivers in the Kingdom of Lesotho in southern Africa. Initially each characterization exercise focused on determining FPZs using an array of remotely sensed data, and this was followed by a ground-truthing exercise where more detailed river channel surveys were undertaken. This further refined the characterization of various rivers and allowed an investigation of the presence and character of functional units and mesohabitats within each zone. The FPZs described for both examples are representative of these rivers and may or may not be present in other rivers globally.

Example 1: Rivers within the Murray–Darling Basin

The rivers of the Murray–Darling Basin in southeastern Australia (Fig. 4.4) have been described based upon the categories discussed in the previous section and shown in Fig. 4.3. Eight FPZs were identified (Fig. 4.4); a description of each is provided in the following sections. Each FPZ has a distinct 'valley floor trough to river' association in terms of the degree of valley confinement, gradient, hydrological and sediment transport regime, stream power (or energy), boundary material or sediment yield, and river channel planform and cross-sectional character. All eight FPZs do not always occur within the individual subcatchments of the Murray–Darling Basin; some may be missing, while others are repeated along a particular river system (see Fig. 4.4). Each FPZ contains a different assemblage of functional sets, functional units, and mesohabitats within the riverscape and the floodscape (Thoms *et al.*, 2004). Thus, a complex array of physical structures is present within the rivers of the Murray–Darling Basin.

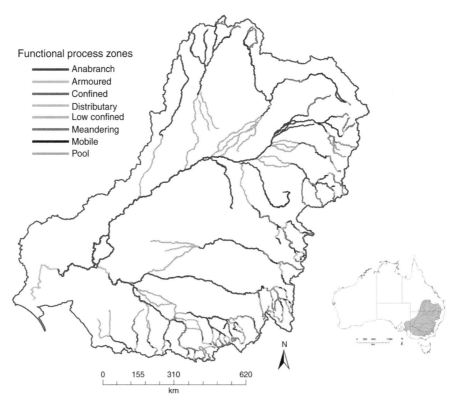

FIGURE 4.4   The spatial distribution of FPZs within the Murray–Darling Basin, Australia, at the 1:500,000 scale (modified from Thoms *et al.*, 2007). Inset shows rivers of Murray–Darling Basin and location in Australia. (See color plate 4).

The physical diversity of river types is associated with a complex spatial arrangement of associated biological communities (Sheldon and Thoms, 2004; Boys and Thoms, 2006). The character of the individual zones is described below.

Pool Functional Process Zone

Pool FPZs are characterized by long, slow-flowing pool sections separated by short higher-energy channel constrictions (Fig. 4.5). The pool sections that form upstream of these channel constrictions are the dominant physical features in these zones and resemble features commonly associated with backwater regions. Channel constrictions are generally associated with major bedrock bars that extend across the channel or substantial localized gravel deposits that act as riffle areas. Local riverbed slopes increase significantly at these constrictions and are small areas of relatively high energy, contrasting to the relatively low bed slopes and lower flow energies of the pool environment. The planform or channel configuration of the pool reaches is controlled by the valley morphology. Pool zones generally have a small flanking floodplain because of narrow valley floor troughs, thereby limiting floodplain development. The bankfull channel can have dimensions of up to 30 m width and 3–4 m depth with width-to-depth ratios of 10. The nature of riverbed substrate in these zones generally consists of fine silt or clay-sized particles, which overly a bedrock or cobble base in the pools. However, gravel and cobble-sized particles and/or bedrock dominate the shorter channel constrictions. During periods of bankfull flows, the finer riverbed substrate is mobilized and actively transported, but discharges greater than $50\,m^3\,s^{-1}$ are required to initiate the motion of the coarser riverbed substrate. Overall, the riverbed sediments in the pool zones are highly stable as indicated by a relative bed stability index between 1 and 2.3. The relative bed stability index is defined as a ratio of the critical velocity required to move a particle and the actual or predicted water velocity near the riverbed. Values less than 1 indicate that particles would be expected to move, and hence that the riverbed would be considered unstable. The suitability of the riverbed for in-stream flora and fauna depends, in part, on the stability of the substrate.

The functional set habitat is the pool zone itself, and the main functional units are the riffle/chute areas in the main channel constrictions and the larger pool areas. Riffles or chutes provide relatively fast-flowing, shallow, turbulent water, while water in pool regions is deeper and slow-flowing. The major mesohabitats are associated with various substrate types, namely cobble or gravel and sand. Mesohabitats appear to be somewhat more diverse within the actual pool areas and have areas of varying substrate combined with emergent and submerged aquatic vegetation.

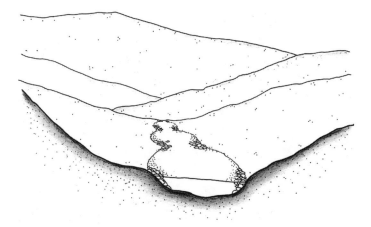

FIGURE 4.5 A schematic of the physical character of a typical pool FPZ.

Gorge or Constrained Functional Process Zones

The gorge or constrained FPZs are relatively high-energy sections of river dominated by steep bed slopes, often greater than 0.010, and with bankfull channel stream powers (the rate of work or potential energy of a length of river) exceeding 400 ωm². Bedrock chutes, large accumulations of boulders or cobbles, and scour pools dominate the in-channel environment. Relatively immobile boulder materials are present; however, floods move any cobble accumulations that are present. These zones lack adjacent floodplains, so sediments are added directly to the channel from adjacent valley slopes (Fig. 4.6). These zones also lack major sedimentary deposits, and this, together with the high flow energy environment, suggests that these areas are an important source zone of sediment for the downstream river system. The riverbed is relatively unstable in constrained zones, with a typical relative bed stability index of 0.46. Channel planforms are controlled by the structure of the valley.

In this zone, the functional set is the constrained section itself. There are few functional units in the constrained zone of the river system. The main channel is the dominant unit, with perhaps some differentiation of riffle areas and pool areas within the channel, but these areas are not as distinct as in the pool zone. Large boulder/cobble accumulations dominate the in-channel environment, and some provide habitat for riparian vegetation. An array of cobble and gravel accumulations generally provides a complex suite of mesohabitats or substrate habitats. However, stands of riparian vegetation also provide complex habitat, both in themselves and because of their associated fallen timber.

Armored Functional Process Zone

Armored FPZs are also relatively high-energy zones with bankfull stream powers often reaching 400 ωm⁻² and have relatively steep bed slopes that range between 0.01 and 0.002. Armoring refers to the development of a surface layer that is coarser than the sediment beneath it. This coarse layer protects the finer materials underneath, and they are not mobilized by the flow until the armored layer is disturbed. Armored zones can also be characterized by a series of floodplains of different ages that are inset into higher-level terraces. These zones can be source areas for sediment, as is shown by the terrace formations and the active erosion of the river banks (Fig. 4.7). The river channel is relatively constrained, being controlled by the width of the valley floor trough and the configuration of the valley. The presence and the extent of adjacent

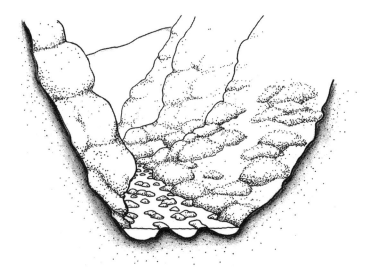

FIGURE 4.6   A schematic of the physical character of a typical gorge or constrained FPZ.

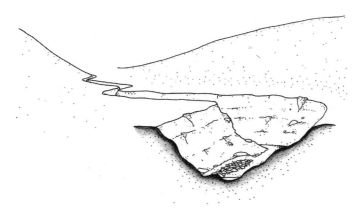

FIGURE 4.7   A schematic of the physical character of a typical armored FPZ.

alluvial surfaces (floodplains and terraces) determine the degree of mobility of the actual river channel. The river channel in the armored zone often has a meandering pattern that is superimposed on a larger valley pattern and is one that is characteristic of a bed load/mixed load channel. The latter features relatively steep bed slopes, little sinuosity, and large meander arcs and wavelengths. The in-channel environment is dominated by cobble and gravel-sized sediments that are extensively armored and relatively stable.

The functional set is the armored zone itself. The armored zone of a river marks the beginning of where the functional units begin to be divided into those occurring in the low-flow channel and those within a high-flow channel. Within the low-flow section of the armored zone, the functional units consist of riffle and pool areas within the main channel. Some of the riffle sections are large and support well-established stands of riparian vegetation. Pool regions are generally large and deep and would provide a substantial refuge area during floods. In this zone, the high-flow channel is present but not well-developed. Functional units in this region of the channel include flat surfaces within the incised channel and the small flood runners. Mesohabitats within the low-flow channel consist of various accumulations of cobble, gravel, and sand particles within the riffles. These often provide an array of complex substrate habitats. Riparian vegetation and snags also provide complex habitat. Large debris dams are often associated with the snags, and these are rich in organic matter. Within the pool unit of the low-flow channel, mesohabitats include emergent and possibly submerged vegetation along with some woody debris. The high-flow channel usually contains sparse mesohabitats, the most dominant of which would be submerged terrestrial vegetation in times of flood.

Mobile Functional Process Zone

A mobile FPZ is an area characterized by a relatively active river channel, as shown by the relatively mobile riverbed sediment and large quantities of sediment in temporary storage within the main active channel (Fig. 4.8). The presence of well-developed inset floodplain features (such as benches) along with point bar systems, cutoffs, and levees testifies to the relatively active and unrestricted nature of this river–floodplain environment. The valley floor is generally wider in this zone, thereby allowing floodplain development to occur. Often the river channel is freely meandering with irregular planforms. Characteristics of the mobile zone are increased meander wavelengths (<2 km) and meander arcs. Stream powers may range from 8 to 20 ωm^{-2}. The morphology of the in-channel environment is extremely variable; bars (point and lateral), benches (at various levels), and riffle/pool sequences may be present. These in-channel storage features reflect high rates of sediment transport. Riverbed sediments typically have a bimodal grain size distribution (median grain size of 64–100 mm) and are highly mobile as noted by a relative bed stability index of <1.

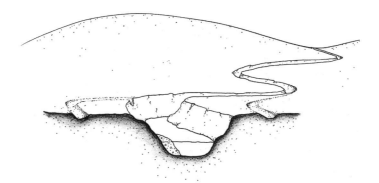

FIGURE 4.8   A schematic of the physical character of a typical mobile FPZ.

The functional set is the mobile zone of the river system. The mobile zone is probably the most complex zone in terms of functional unit development, with distinct and diverse low-and high-flow channels. Within the low-flow section, functional units are again the riffle and pool areas within the main channel. Generally, the riffle sections are large and support well-established stands of riparian vegetation, while the pools are also large and deep. The low-flow channel is also characterized by large sandy point bars. In the mobile zone, the high-flow channel is normally well-developed with in-channel benches, diverse flood runners, and an extensive floodplain. Functional units in the high-flow channel include the flat bench surfaces, the small flood runners, and complex features of the floodplain itself. Mesohabitats within the low-flow channel again depend on substrate composition, with accumulations of cobble, gravel, and sand within the riffles providing complex substrate habitats. The stands of riparian vegetation also provide complex habitat, both themselves and their associated fallen timber. These large woods create debris dams, much as they do in the armored zone. Within the pool functional unit of the low-flow channel, microhabitats include emergent vegetation. Submerged vegetation is typically not as abundant in this reach owing to the overall depth of the channel. Woody debris is also present within pools. Mesohabitats within the high-flow channel depend on inundation of the floodscape during floods; however, snags and fallen woody debris form a major microhabitat in this region of the channel.

Meander Functional Process Zone

A distinguishing feature of the meander FPZ is the significant increase in the width of the valley floor (generally >15 km wide) and the associated floodplain surface. The river channel is relatively active and displays a typical meandering style. Sinuosities may range from 1.8 to 2.35 with meander wavelengths between 200 and 700 m. The presence of well-developed floodplain features, such as flood channels, former channels, avulsions, cutoffs, and minor anabranching, testifies to the relatively active and unrestricted nature of the river–floodplain environment in this reach. The river in this zone is typical of a mixed to wash load channel. The morphology of the in-channel environment is variable with the presence of bars, benches, and riffle/pool sequences (Fig. 4.9). These in-channel sediment storage features reflect the relatively high rates of sediment transport. The riverbed sediments typically have a bimodal grain size distribution like those found in the mobile zone, but are generally smaller in size (median grain size from <1 to 64 mm). The appreciable fining of the bed sediment makes a clear distinction between the meander and mobile zones. The bank sediment is also very fine, with mostly fine sands, silts, and clays. The cohesive nature of the bank sediments contributes to relatively steeper banks in this zone compared to zones elsewhere in the river network.

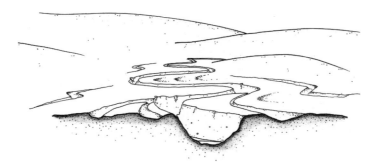

FIGURE 4.9   A schematic of the physical character of a typical meander FPZ.

The functional set is the meander zone of the river system. There is a complex array of functional units within both the low- and high-flow channel units. Riffle and pool areas dominate the main channel. Riffles can be especially large and support well-established stands of riparian vegetation. Pool areas are large and deep. The low-flow channel is also characterized by large sandy point bars. In this zone, high-flow functional units are well-developed, and in-channel benches, diverse flood runners, and an extensive floodplain are all present. Functional units include the flat bench surfaces, the small flood runners, and complex features of the floodscape itself. Mesohabitats include the complex array of substrate types, stands of riparian vegetation, snags, and debris dams. Within the pool functional unit of the low-flow channel, mesohabitats also include emergent and submerged vegetation, although this is generally not abundant owing to the overall depth of the channel.

Anabranching Functional Process Zone

The channel in this FPZ often flows across a very broad, low-angle fan complex or outwash surface and is commonly associated with a series of effluent channels or anabranches (Fig. 4.10), which is typical of many of the rivers within the Murray–Darling Basin. The main river channel is characteristic of a wash load system with relatively low bed slopes, high sinuosities, low bankfull stream powers, and highly cohesive bank materials. River channels in this zone can be described as being under fitted within an older large palaeochannel system. Usually the contemporary channel has a sinuosity of 2.06 or less, which is contained within the older channel system that has a much larger meander wavelength and channel dimensions. The active channel can have bankfull characteristics of low width/depth ratios, widths of 30–150 m, and depths of 10–15 m with a highly stable riverbed (relative bed stability index of 0.98–4.85). Anabranch channels begin to flow at approximately one-third to one-half bankfull discharges; therefore, bankfull capacities in this zone are lower in comparison to other FPZs through the river network.

FIGURE 4.10   A schematic of the physical character of a typical anabranching FPZ.

The functional set is the anabranching zone of a river system. The anabranch zone is typical of many lowland rivers in western New South Wales. In this zone, the low-flow channel is relatively simple, with most of the habitat diversity occurring at higher-flow levels. Within the low-flow section of the anabranching zone, riffle functional units do not exist and the main functional units are large pools within the main channel. Sections of the low-flow channel may also be characterized by large sandy point bars. By comparison, the high-flow channel is also well-developed with in-channel benches occurring at various levels within the channel, diverse flood runners, and large anabranches leaving the main channel at various flow heights, and an extensive floodscape. The major mesohabitats depend on the sediment composition of the riverbed substratum; however, geomorphic diversity is limited to sandy bars and regions of silt/clay. Woody debris from fallen riparian vegetation is the other major mesohabitat of the low-flow channel. Mesohabitats within the high-flow channel are similar to those within the low-flow channel with woody debris dominating. At extremely high flows, mesohabitats are associated with inundation of the floodscape; however, snags and fallen woody debris again form a major microhabitat in this region of the channel.

Distributary Functional Process Zone

A series of bifurcating channels (channels that take off from each other) are the main distinguishing features of distributary zones (Fig. 4.11). These secondary channels persist relatively independently of the main channel for lengths far in excess of their width. Distributary channels may rejoin the main channel or each other, and a feature of these channels is a *decrease* in bankfull cross-sectional area or channel size downstream. The Namoi distributary zone, for example, commences at a well-defined north–south topographic boundary slightly east of Narrabri. In all channels, there is a decrease in bankfull cross-sectional area or channel size downstream. This is attributed to the loss of water by evaporation and flood attenuation. Sediments within all the channels are composed of very fine sand, silt, and clay-sized particles. Indeed, the percentage weight of silts and clays can be up to 50%. Most of the channels are relatively narrow and featureless, with occasional deep holes scattered along their length.

There are relatively few major habitats within the channels of this FPZ because the low-flow channel is relatively simple in cross-sectional shape. Most of the functional units occur within the higher-flow channels. Within the low-flow section of the distributary zone, large deep pools are common. Sections of the low-flow channel may have the occasional point bar. In the high-flow section of the main channel, in-channel benches occur at various levels, and these are considered important for in-stream ecological processes. The secondary channels and the extensive floodscape through which they flow are the dominant functional units in this zone. The high-flow channels also contain flat bench surfaces, flood runners, and anabranches and a complex array of floodplain features. Woody debris from fallen riparian vegetation is probably the major mesohabitat of the low-flow channel. Mesohabitats within the high-flow channel are

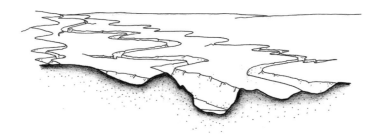

FIGURE 4.11   A schematic of the physical character of a typical distributary FPZ.

similar to those within the low-flow channel, with woody debris dominating. At high flows, mesohabitats relate to the secondary channels and inundated terrestrial environments; however, snags and fallen woody debris again form a major mesohabitat in this region of the channel.

In summary, the anabranch zone is the dominant FPZ in the Murray–Darling Basin (Fig. 4.4), comprising 39% of the total length of rivers mapped (2.8 million km), while the pool FPZ is the rarest FPZ, accounting for <1% of the total river length. None of the subcatchments contained all FPZs (Fig. 4.4). A more detailed census of the composition of FPZs and the implications for river management is provided in Chapter 8.

Example 2: The Rivers of the Kingdom of Lesotho

River systems in the Kingdom of Lesotho are highly constrained within deep valleys, an artifact of the region's geological history (Partridge and Maud, 1997). Continental disturbance since the Mesozoic resulted in a series of uplift events in southern Africa. This was associated with extensive river incision. Regional variations in geological structure, rates of uplift, drainage network evolution, and climate have produced an array of valley floor trough–river complexes. The important controls on the morphology of the associated river channel relate to the nature of the valley floor trough, in particular, its size relative to the river channel, configuration, angle of adjacent valley slopes, gradient, geology, and the texture of sediments contained within the trough. For example, increasing valley floor width relative to the main river channel provides areas for sediment deposition and construction of floodscape areas. This is important because the presence of floodplain surfaces isolates the river channel from direct sediment additions from adjacent hill slopes. Spatial variations in valley floor trough conditions have been shown to influence flood hydraulics (Miller, 1995), sediment transport and deposition (Thoms et al., 1998), and channel morphology (Warner, 1987).

Six main valley floor trough–river associations occur in Lesotho. Each has its own distinct set of river channel morphologies, and these do not always occur in a clinal pattern or in order; some may be missing while others may be repeated along the major river systems of the Senqu, Senquyane, Malibamatso, and Matsuko (Fig. 4.12). As a complex, there is an array of physical structures present along the length of individual rivers. Each valley floor trough–river association represents a distinct FPZ reflecting changes in the degree of valley dimensions, gradient, stream power (or energy), boundary material, and sediment yield. These differences influence the river's ability to adjust to changes in flow and sediment transport regimes. Moreover, as individual zones have a different physical character, this will have implications on the presence and structure of instream habitat and associated biological communities. The character of each zone is described below.

Headwater Functional Process Zone

The river channel in this zone occupies the entire valley floor trough, and as such its planform configuration is heavily influenced by valley conditions. Located in the upper most regions of all the main river systems in Lesotho, they are characterized by a wide range of poorly sorted riverbed sediments that are added directly to the channel from adjacent valley sides. Essentially, the physical structure of the riverscape is heavily influenced by the surrounding bedrock. Minor accumulations of sediments occur locally at flow expansion zones or upstream of flow constrictions. These deposits serve as important aquatic refugia during the dry winter months because sections of this zone are ephemeral. Their location is dependent upon catchment size and rainfall.

Few functional units exist in the headwater FPZ of these river systems. The main channel is the dominant unit with perhaps some differentiation of chute/rapid and pool areas within the channel. Present within the main channel are bedrock sills and ledges along with large boulder/cobble

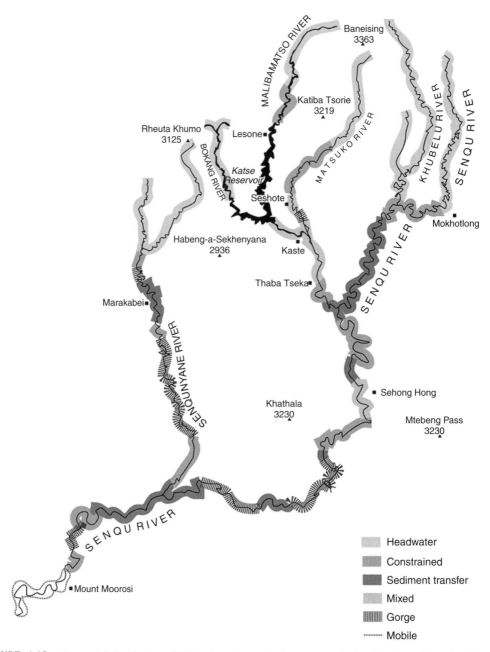

FIGURE 4.12 The spatial distribution of FPZs along the main river systems in the Kingdom of Lesotho. Note the repeating FPZs, particularly the gorge, sediment transfer, and mixed FPZs.

accumulations. Within the channel, the major mesohabitats relate to substrate composition with bedrock and accumulations of cobble/gravel providing complex substrate habitats.

Gorge Functional Process Zone

This zone resembles a typical canyon river system (Schumm, 1988) with near vertical valley sides and a river channel that occupies the entire valley floor during all flows. As a result, a range of materials are added directly to the channel. This is a high-energy zone because of narrow floor widths and high bed slopes ($S > 0.010$); hence bankfull channel stream powers can exceed 5000 ωm^2. Channel planforms are controlled by the structure of the valley. Bedrock chutes, rapids, and runs along with large boulder/cobble accumulations and scour pools dominate the in-channel environment. Boulder materials are relatively immobile; however, cobble accumulations are probably highly mobile and unstable with an estimated relative bed stability index of 0.46 during flood flows. Major sedimentary deposits are lacking in this zone because of high stream powers. Thus, this zone is an important sediment source zone because of the high energies and the direct supply of sediment from the catchment. Functional units are rare in the gorge zone of these river systems. The main channel itself is the dominant unit with perhaps some differentiation of chute/rapid and pool areas within the channel. Dominant within the channel are large boulder/cobble accumulations, some of which may provide habitat for riparian vegetation. At local valley expansion areas, lateral bars would be temporary habitats. Within the channel functional unit, the major mesohabitats relate to substrate composition, with bedrock and accumulations of cobble/gravel providing complex substrate habitats.

Constrained Functional Process Zone

Decreases in gradient, lower valley slopes, and relative increases in valley floor widths are important features of this river zone, which distinguish it from the gorge zone. Hence, the direct influence of adjacent valley slopes is decreased, especially in terms of sediment supply even though the river channel occupies the entire valley floor trough. Regular in-channel accumulations of sediments occur in this zone because of lower sediment transport efficiencies. Lower sediment conveyance results from lower slopes and relative increases in valley floor widths. Four main types of sediment bars are present in various abundances, with each being small in comparison to the size of the river channel: (1–2) lateral bars, both mid-channel and those attached to channel margins; (3) point bars; and, (4) tributary mouth bars, which form downstream of tributary junctions. Bedrock has a variable influence on the structure of the river channel in this zone. For example, at some locations, bedrock outcrops form the loci for bar development, while at other sites, they form riffles or chutes. The riverbed sediments are highly armored at locations dominated by cobble, gravel, and sand.

The river channel and the major sediment deposits contained within the active channel are the dominant units in this zone. The river channel itself has riffle/pool sequences at irregular intervals because of the variable bedrock influence. Chutes and rapids are also important units, especially at local bedrock constrictions. These are not as dominant in size or number in comparison to those in the gorge zone. In-channel bars are an important functional unit in the constrained zone, and each bar type (lateral, point, or tributary) has a unique physical structure. Within the channel unit, the major mesohabitats relate to the highly variable substrate composition. Bedrock surfaces dominate part of the river margins, but small boulders and accumulations of cobble/pebble/gravel also add to the complexity of in-channel substrate habitats. Sand deposits are also present in low-flow zones, such as the lee of boulder and cobble/pebble accumulations.

Sediment Transfer Functional Process Zone

The bankfull or active riverscape in this FPZ does not occupy the entire valley floor trough. The relative increase in the width of the valley floor trough and slight decreases in river slopes facilitate deposition of significant quantities of sediment for various time periods. The presence of large volumes of alluvium is a dominant feature of this zone, contrasting with that of headwater, gorge, and constrained zones. Terraces, floodplains, in-channel benches, and numerous bar deposits are present in this river zone. The lower surfaces (inset floodplains and in-channel benches and bars) obviously have shorter residence times, as displayed by active erosion of these deposits. They are, therefore, both important sediment storage and source areas.

Large sediment bars are a dominant feature of the in-channel and can often occupy up to 75% of the active river channel. They are highly mobile as displayed by the lack of sedimentary structures contained within the deposit, the presence of low-angle bed forms across the surface of the bars, and the overall morphology and functional nature. Five main bar forms were observed in this zone, as described below.

- Longitudinal bars, which form in the center of the channel at relatively wide locations. They are convex and elongated. They grow by upward accumulation of coarser material and downstream accumulation of finer material. These bars are closely associated with crescentic bars, which Bluck (1982) suggested are highly mobile and youthful forms of longitudinal bars.
- Transverse bars tend to occur at abrupt channel expansions or downstream of large tributaries. They often exhibit broadly lobate or sinuous fronts and a steep downstream face or ramp – this distal face is often referred to as an avalanche face. Generally the main flow moves over the center of the bar and therefore they may be centrally concave.
- Point bars occur near the convex bank of curved channels.
- Diagonal bars are singular in that they are oriented obliquely across the channel rather than parallel to the flow. Often they are attached to the bank at both sides.
- Tributary bars form at tributary junctions in a large fan shape and resemble transverse bars. Often there is a degree of symmetry of the fan. This is dependent on the geometry of the tributary junction and the relative character of the two contributory streams.

Longitudinal, crescentic, and transverse bars were the most common in this zone. These forms are considered to be the most unstable form of channel bar (Bluck, 1982). The presence of bars in this zone had the effect of dividing the channel into multiflow paths.

This FPZ is probably the most complex in terms of the assemblage of functional units present with distinct and diverse low- and high-flow channels. Within the low-flow section of the sediment transfer zone, functional units are various bar structures, and the small flow channels cut into them. The river channel itself had regular riffle/pool sequences, although this regularity is often disrupted because of variable bedrock influences. Pool regions are also large and deep. Rapids, chutes, runs, and scour pools are also present. In the sediment transfer zone, a high-flow channel is present, which has well-developed in-channel benches, diverse but small flood runners, and several floodplain surfaces. Functional units in the high-flow channel include the flat bench surfaces, the small flood runners, and complex features of the floodplain itself. Within the channel unit, the major mesohabitats relate to the highly variable substrate composition. The presence of bedrock surfaces is variable and generally restricted to the river margins, while a mixture of cobble, pebble, gravel, and sand material is dominant and added to the overall complexity of in-channel substrate habitats.

Mixed Functional Process Zone

No dominant river channel structure is present in this zone. Although other river zones identified in Lesotho have a sufficient length of river to be recognized, the mixed zone does not. Mixed zones are sections of river where there are highly variable regional valley floor trough conditions. Valley floor dimensions change irregularly between highly constrained and sediment transfer sections. Individual subsections may be up to 1–2 km in length. Mixed zones are then highly variable tracts of river with complex arrays of physical structures.

The only distinguishing feature of the mixed zone was the relative increase in the contribution of sand in the riverbed and floodplain sediment. This zone is complex in terms of functional unit development. There are both distinct and low- and high-flow channels. Within the low-flow section of the mixed zone, functional units are various bar structures, and the small-flow channels cut into them along with bedrock surfaces. The river channel itself has regular riffle/pool sequences although this regularity is disrupted because of variable bedrock influences. Rapids, chutes, runs, and scour pools are also present. There are high-flow functional units, and these are associated with those sections of river where valley floor troughs increase in width. In-channel benches are present, and these have small flood runners eroded into them. Other high-flow functional units include an array of floodplain surfaces at various heights above the main channel. The major mesohabitats again relate to the highly variable substrate composition. Bedrock surfaces are variable and present both within the channel and at the margins of the channel. Sand dominated the in-channel mesohabitats although a mixture of cobble, pebble, and gravel material is also present.

Mobile Functional Process Zone

Distinguishing features of the mobile zone include significant increase in valley floor width (>5 km) and decrease in river channel gradient. This is associated with a general reduction in topography, as the main rivers of Lesotho debouch from the highlands. The relative increase in valley widths is associated with the presence of well-developed floodplains and terraces. The morphology of the in-channel environment is variable with the presence of bars (point and lateral) and small benches (at various levels). These in-channel sediment storage features reflect the relatively high rates of sediment transport through this zone. The riverbed sediments are dominated by fine gravels and sands. The appreciable fining of the bed sediment is a clear distinguishing feature between the mobile and other zones. The bank sediment is also very fine, mostly fine sands, silts, and clays. The cohesive nature of the bank sediment contributes to relatively steeper banks in this zone compared to upstream zones.

The mobile zone contains a complex array of functional units with distinct and low- and high-flow channel units. Within the low-flow section of the mobile zone, functional units are again the riffle and pool areas within the main channel. Pool regions are also large and deep. The low-flow channel is characterized by large sandy lobate and point bars. In this zone, high-flow functional units are developed with in-channel benches, diverse flood runners, and an extensive floodscape. Functional units include the flat bench surfaces, the small flood runners, and complex features of the floodplain itself. Within the low-flow channel, the major mesohabitats again relate to substrate composition, with accumulations of fine gravel and sand within the riffles providing complex substrate habitats. Limited stands of riparian vegetation also provide complex habitat.

WHAT SCALE TO CHOOSE AND ITS RELEVANCE TO RIVERINE LANDSCAPES

A view that riverine landscapes are an interacting system of physical components, at multiple scales, is a theme pursued throughout this book. The physical character or habitat provides the template upon which evolution acts to forge characteristic strategies of life history (Southwood, 1977). Accordingly, the physical properties of any given habitat within a river ecosystem will influence the type, abundance, and arrangement of biological assemblages found there. Interactions between the biological and physical components of a river ecosystem generate pattern, and it is the goal of river science to decipher the causal mechanisms, or processes, underlying observed patterns (Levin, 1992; Fisher, 1993; Schumm, 1988). The interaction between pattern and process is not unidimensional but, rather, occurs hierarchically across multiple scales (Wiens, 1989; Levin, 1992; Peterson and Parker, 1998). A pattern at one scale may be generated by processes operating at different hierarchical levels. Similarly, a process may be influenced by patterns occurring at multiple scales. The interplay of pattern and process within an ecosystem generates a complex matrix of interactions. A challenge in river science is to dissect the patterns and processes in hierarchical, multicausal ecosystems, in this case riverine landscapes, into spatial and temporal domains of influence.

The extended fluvial hydrosystem approach presented in the previous sections is new on several fronts, and it allows the river scientist to explore pattern and process at multiple scales. First, it introduces the component of FPZs at a spatial scale, which is nested between valley settings and river reaches. This level of organization has been missing from the study of rivers, and its inclusion aids in constructing a more complete, multiscalar view of the riverine landscape. Providing detailed information at multiple levels of organization is critical in interdisciplinary studies because scientists generally only impose scales of observation commensurate with their disciplinary experience of the system in question (Thoms and Parsons, 2003). Individual disciplines view patterns and processes according to their own paradigms and theories, often resulting in mismatches of scales of observation between disciplines. Limiting the levels of organization in any hierarchical system ultimately leads to the limited explanation of system structure and functioning. Second, in the case of riverine landscapes, FPZs often emerge as an appropriate level of investigation in river characterization. Selecting the appropriate levels for investigating pattern and process is important for the study and management of river systems. There are many different ways to approach the issue. It is common to have the scale of investigation preset either directly by the issue at hand or by some management action. Here, organizational levels are predetermined, and the attributes that are likely to characterize different components of the river systems are preselected based on this level of organization. However, in situations where levels of organization or scales have not been predetermined, they may self-emerge (Parsons and Thoms, 2007; Thoms *et al.*, 2007; Thoms and Maddock, unpublished data). Here we offer two examples to illustrate this point. The first examines associations between environmental influences and macroinvertebrate assemblage distribution across a hierarchy of river system organization in the upland Murrumbidgee River catchment of Australia (Parsons *et al.*, 2004b). In this river, Parsons and Thoms (2007) demonstrated that different scaled environmental factors, collected at the catchment, FPZ, reach, and riffle scales and related these to the region and cluster levels of macroinvertebrate distribution. The hierarchical pattern of large, region-level and local, reach-level macroinvertebrate distribution was matched by FPZ scale and local reach scale of environmental influence. Intermediate zone-scale environmental factors and smaller riffle-scale factors were not important influences. Thus, larger regions like FPZs and local reaches were considered important levels of organization for macroinvertebrate–environment associations in rivers of the upper Murrumbidgee catchment. The second study involved an investigation of the distribution of in-channel habitats. Here Thoms and Maddock (in press) used a series of multivariate statistical analyses to assess the distribution of in-channel geomorphic units or physical habitat patches at scales between 50 m and 1 km, enabling the self-emergent properties of scale to be identified. No groupings of in-channel habitat were evident at the 50-m scale, but six habitat groups were apparent at the 100-m scale in the Cotter River, Australia. These groups were dominated by combinations of runs, glides, and pools, but there was no clear spatial dimension to the distribution of these groups along the river. However, a clear spatial dimension to the organization of physical habitat was apparent at the 500-m and 1-km scales where three distinct FPZ emerged, each differentiated by their combinations of in-channel habitats. The ecological relevance of this approach was demonstrated through its application in defining the connectivity and fragmentation of prime habitat for an endangered native fish species, the Macquarie perch (*Macquaria australasica*). This information was used for determining key locations for habitat enhancement, assessing the likelihood of success of reintroducing native species, and assessing the effects of environmental flows on fish distributions.

To further highlight the role of FPZs as an appropriate and convenient level of organization with which to investigate the structure and functioning of riverine landscapes and its ecological significance to riverine landscapes, we illustrate this with an example of the Barwon–Darling River in southeastern Australia. In particular, we consider whether the extended fluvial hydrosystem approach outlined above provides a suitable characterization scheme from which intermediate spatial scales, as provided by FPZs, can be selected for the basis of establishing fish habitat associations in riverine landscapes. Here, the research of Boys and Thoms

(2006), Boys *et al.* (2005), and Sheldon and Thoms (2004) are summarized. The premise of the approach is that the composition of a biological assemblage in an area of river is constrained by the availability of aquatic habitat, which is essentially governed by the geomorphological processes operating within the area. Various empirical studies have demonstrated that river zones of similar physical character can be predicted in smaller streams (Lanka *et al.*, 1987; Walters *et al.*, 2003) and suggest that such zones may be useful spatial strata to consider when examining the distribution of habitat and biological assemblage composition in larger riverine landscapes. By incorporating FPZs into their fish sampling design, Boys and Thoms (2006) demonstrated the utility of this level of organization as it takes into account the relatively large size and spatial variability of dryland rivers, such as the Barwon–Darling River.

Four FPZs have been identified along the Barwon–Darling River between Mungindi and Wentworth (Thoms *et al.*, 2007), all nested within a valley zone characterized by highly variable styles of river channel anabranching. In Zone 1, between Mungindi and Collarenebri, bankfull cross-sectional area increases downstream and the channel has an average bankfull capacity of approximately $460 \, m^3 \, s^{-1}$; this contrasts with Zone 2 (Collarenebri–Brewarrina) where cross-sectional areas decrease downstream. Zones 3 (Brewarrina–Bourke) and 4 (Bourke–Louth) have a relatively constant bankfull area, but the latter is 50% larger. In Zone 4, the lowermost river zone, the channel decreases in size as you move downstream. For the purposes of this study, which was to consider physical habitat fish associations along a 1000 km of river, three 10-km reaches were randomly selected within each of the four FPZs. Fish habitat was mapped along each reach during low-flow conditions. It is pertinent to note that few physical features are available to fish to act as habitat during low flow. The main physical features present, commonly referred to as mesohabitats in the geomorphological and ecological literature, were large wood, three edgewater patches (smooth bank, irregular bank, and matted bank), and two open water patches (mid-channel and deep pools; see Table 4.2). Thus a hierarchical design of mesohabitats nested within reaches, reaches nested within zones, and zones nested within the river network (Boys and Thoms, 2006) was available. To complement the physical habitat survey, fish surveys were conducted during daylight hours at each of the 12 study reaches using boat-mounted electrofishing. Fishing operations were performed at each of the six mesohabitat types within the 12 reaches. The location of these operations within each reach was selected at random depending on the availability of each mesohabitat type. Details of the methods used are provided by Boys and Thoms (2006) and Boys *et al.* (2005).

Some 5526 fish were recorded from all sites in the Barwon–Darling River. Of the total abundance, 86% were native to the area and 14% were alien species. Most of the native fish

TABLE 4.2 Physical Structures Available to Fish for Habitat During Low-Flow Conditions in the Barwon–Darling River

Habitat attribute	Description
Smooth bank	Uniform sedimentary bank structure contacting the waterline. Identified as percent bank cover of 100 m of channel
Matted bank	Fine matted structure (root mats or lignum) contacting the waterline. Identified as percent bank cover of 100 m of channel
Irregular bank	Complexity at the waterline caused by rotational shear of the bank or rock outcrops. Identified as percent bank cover of 100 m of channel
Structural woody habitat	Wood structure along the bank, contacting the water line. Identified as a tally per 100 m of channel
Open channel	Depths shallower than the 70th percentile. Measured by depth sound every 100 m along the thalweg
Deep water	Depths deeper than the 70th percentile. Measured by depth sound every 100 m along the thalweg

were bony herring (*Nematalosa erebi*; $n = 4529$), with reasonable catches of golden perch (*Macquaria ambigua*; $n = 152$). Relatively few Murray cod (*Maccullochella peelii peelii*; $n = 39$), freshwater catfish (*Tandanus tandanus*; $n = 2$), gudgeons (*Hypseleotris* spp.; $n = 5$), and the threatened silver perch (*Bidyanus bidyanus*: $n = 2$) were caught. The common carp (*Cyprinus carpio*; $n = 781$) was the predominant alien species caught along with goldfish (*Carassius auratus;* $n = 15$) and mosquito fish (*Gambusia holbrooki*; $n = 1$).

A distinct longitudinal distribution of fish assemblage was noted in the Barwon–Darling River, and this change in assemblage structure closely corresponded to the four FPZs. The main findings were the following:

- *Zone 1*: Multivariate analysis of the fish data demonstrated a clear separation between FPZs 1, 2, and 3 (Fig. 4.13), a finding supported by an analysis of similarity.
- *Zone 2*: The fish assemblage of this FPZ is distinctly different in composition to all other FPZs (Boys and Thoms, 2006); indeed, a significant difference was detected in the composition of the fish assemblage between Zone 2 and all other zones. This difference in composition is largely attributable to the lower abundance of native species (bony herring, golden perch and Murray cod) and higher abundance of alien species (carp and goldfish).
- *Zone 3*: The three reaches of this FPZ formed a significantly distinct group in ordination space (see Fig. 4.13). Significant differences in the composition of the fish assemblage of this FPZ and the two upstream zones are notable. Species contributing most to these differences were bony herring, carp, and golden perch, with bony herring and golden perch being more abundant in Zone 3 than the other two zones. The Brewarrina to Bourke zone also contained more carp than Zone 1 but similar numbers of carp to Zone 2.
- *Zone 4*: There was a large amount of variation in the composition of the fish assemblage within Zone 4. This variation is attributable partly to the low numbers of fish recorded from reach 11 when compared to reaches 10 and 12. Reach 12 (the furthest downstream reach) had the highest species richness of all the reaches, with freshwater catfish, silver perch, and gudgeon all recorded. Because these species were recorded only at one of the three reaches within the zone, they cannot be reliably used to discriminate between Zone 4 and the other zones.

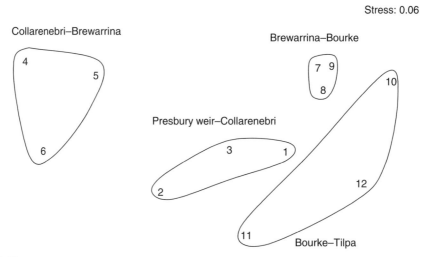

FIGURE 4.13 Two-dimensional, multidimensional scaling ordination of Barwon–Darling River sites based on similarities between fish assemblages. Reaches are grouped by FPZ (from Boys and Thoms, 2006).

This study of fish–habitat associations has demonstrated that the physical heterogeneity of the Barwon–Darling riverine landscape is reflected in a spatial biological heterogeneity. Distinct regional differences in the composition of the fish assemblage along the Barwon–Darling River were recorded. These regional differences involve changes in the relative abundance of species rather than the addition or replacement of species. Two bioregions of particular note within the Barwon–Darling River were identified at a spatial scale that closely corresponded to the zone level of geomorphological classification used. The fish assemblage of Zone 2 differed from that of all other zones and contained a larger proportion of carp (an introduced species) than the other zones, with significantly fewer bony herring, golden perch, and Murray cod. In comparison, Zone 3 immediately downstream had a fish assemblage characterized by comparatively higher abundances of the native species such as bony herring, golden perch, and Murray cod. There are a few scenarios that may explain this river-scale spatial trend observed in the fish assemblage. Zone 2 may be composed of less-desirable mesohabitat types than Zone 3. Based on the patterns of mesohabitat association observed, this may include a lower density of large wood. Alternatively, the presence of a significant impediment to fish passage during low flows (the Brewarrina Weir) may be preventing the recolonization of habitats in Zone 2 by downstream fish populations. Another alternative involves the proximity of Zone 2 to tributaries that are known to be hotspots for carp recruitment, namely the Namoi and Macquarie Rivers. All three scenarios probably exerted some degree of influence over the spatial character of the fish assemblage at the river scale.

Describing regional differences in biological communities along entire riverine landscapes is not new. Longitudinal fish zonation concepts (Huet, 1954), for example, have been used for many decades in Europe (Aarts *et al.*, 2004), and refined versions still play an important role in present-day riverine fish surveys. A fundamental drawback of any biological zonation concept is that data are typically obtained from the random sampling of reaches. Although they may describe longitudinal changes in biological assemblage composition, they do little to explain why such changes may occur. Longitudinal gradients of change in a river may merely be an artifact of positive spatial autocorrelation (Legendre, 1993), a phenomenon frequently encountered in assemblage data whereby nearby points in space are more similar than would be expected by chance. The reason for this may be of an environmental (e.g., gradients of temperature or physical habitat change) or biological (e.g., the dispersal ability of species within the assemblage) nature. By observing biological assemblage at particular reaches in the context of both a mesoscale (a scale below in the hierarchy) and the FPZ scale (a scale above in the hierarchy), the approach outlined here by Boys and Thoms (2006), and also strongly advocated and developed in this chapter, allows us to test whether the spatial autocorrelation of reaches may be due a homogeneity of habitats resulting from and constrained by the influence of geomorphological processes occurring at both smaller and larger scales. Such insight can be achieved only after analyzing the results of habitat surveys conducted at multiple scales in a hierarchy and, in particular, through an emphasis on FPZ-scale analyses.

SUMMARY

Many of the traditional paradigms that modeled rivers as one-dimensional and equilibrial systems (e.g., the RCC, Vannote *et al.*, 1980) are being replaced by conceptual models and frameworks that emphasize the importance of scale, hierarchy, complexity, variability, heterogeneity, and stochasticity (e.g., Thorp *et al.*, 2006; Dollar *et al.*, 2007; Parsons and Thoms, 2007). Although there is a general consensus that these factors are important characteristics of riverine landscapes, relatively few empirical studies explicitly include these within their study designs or conceptual frameworks. On the contrary, they are often removed to eliminate or reduce confounded statistical designs and interpretations. It is apparent that in many aspects of their physical and biological character, riverine landscapes are composed of a mosaic of patches. At the catchment scale, FPZs have emerged as an important level of organization or scale that should be recognized and included within the hierarchical composition of riverine landscapes. These river zones represent larger-scale patches of the riverine ecosystem that have similar flow and sediment regimes and, as a result, comparable channel and floodplain morphologies. Functional process zones do not display a longitudinal continua, rather they comprise a patch mosaic at the catchment scale and can repeat themselves within the riverine ecosystem. Awareness and understanding of this level of spatial organization within riverine landscapes are crucial for river scientists.

Understanding the spatial organization of complex systems is a cornerstone for their study, conservation, rehabilitation, and management. Identifying and modeling the spatial dynamics of environmental systems, at multiple scales, enable us to identify the nature, location, and rate of change as well as likely trajectories. Without this understanding, we cannot comprehend the likely impact of a disturbance(s). This is especially important where sustainable solutions are required to deliver ecosystem services vital for human survival and development. Acquiring this knowledge for riverine landscapes can be a daunting task because of the number of interacting variables operating across multiple scales, which often occur in a stochastic manner. The nonuniform, heterogeneous, and multiscaled nature of riverine landscapes is daunting not only to river scientists but also to policymakers and decision makers, many of whom seek uniformity and simplicity. The first step in the quest for sustainable approaches to the study and management of these ecosystems is the explicit recognition of their spatial organization and diversity and the nonequilibrial nature of interactions occurring within them (Kay *et al.*, 1999; Parsons and Thoms, 2007).

There are two aspects to enhancing our spatial awareness of river landscapes. First, to overcome issues of scale, it is important to have data or information for as many scales as possible and then allow the scale(s) of systems behavior and character to self-emerge rather than to impose a scale for their study or management. Thus we allow spatial considerations to be intimately linked to our knowledge of riverine landscape dynamics over multiple scales. For this, nested hierarchical frameworks is an elegant tool with which to organize information (cf., Dollar *et al.*, 2007) and is the one that provides a basis for entire riverine landscapes to be studied or managed. Second, rapid developments in technology have resulted in the increased availability of spatial data at larger scales. In particular, the use of remotely sensed and real-time data has increased brought about an awakening in spatial awareness, with increased appreciation of larger scales within an increasingly connected world.

Defining the Hydrogeomorphic Character of a Riverine Ecosystem

INTRODUCTION

Riverine landscapes are a mosaic of patches (Thoms *et al.*, 2005). A *patch* can be defined as a relatively homogenous area, in space and time, which differs from its surroundings and other patches (Forman, 1997). They may differ in size, shape, type, and degree of heterogeneity. Patches within the riverine landscape mosaic can be determined through various approaches and methods. The ordering of sets of observations or characteristics into meaningful groups or patches, based on their similarities or differences, is common in river science. However, patches can also be distinguished according to discontinuities in character from their surroundings and from one another. This latter approach focuses on identifying boundaries between groups rather than looking for common groups. Whatever be the approach, the process of patch recognition within a riverine landscape is scale-dependent. A patch at any given scale has an internal structure that reflects its patchiness at finer scales (Kotliar and Wiens, 1990). A patch at the scale of a mesohabitat would typically be represented by clusters of different sediment textures, such as sand and silt-sized particles, whereas an amalgam of functional units such as riffles, pools, and channel bars represent different patches at the river reach scale. At the scale of a river network, FPZs or river types would represent a patch. Collectively, these physical features represent a hierarchy of patches as postulated by Wu and Loucks (1995) in their concept of HPD.

The diversity of river forms and physical structures present at various scales provides significant challenges for the development of flexible and generic approaches to river classification. In terms of river networks, river scientists are required to identify the characteristic forms and features along with the possible processes responsible for producing and maintaining them. Through a process of classifying the physical character of river systems, knowledge of the processes responsible for their formation can be gained. This is based on the premise that rivers are process-response systems. As such, their morphology (physical form) is influenced by the interaction of a suite of independent variables that set the boundary conditions of the riverine landscape and a set of process variables (flow and sediment transport) that sculpture the riverine environment. The literature is replete with studies demonstrating relationships between discharge and/or sediment transport (important physical processes) and the morphology (the physical form) of river systems. A generalized relationship for stable river channels was proposed by Lane (1955). It depicted an interaction between sediment discharge (Q_s), stream discharge (Q_w),

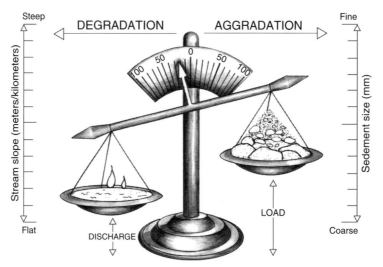

FIGURE 5.1   A generalized relationship of a "stable channel" showing the balance between discharge, slope, particle size, and sediment load. Changes in any of these variables results in a change in the morphology of a river. After Lane (1955).

particle size (D_{50}), and slope (S), whereby a change in any of these variables initiates a series of mutual adjustments resulting in a direct change in the morphology of the river. This generalized relationship is shown in Fig. 5.1. Relationships of this type have dominated much of the fluvial geomorphology literature for many years and have included the establishment of empirical relationships between discharge and bankfull width and depth (Leopold and Wolman, 1957), bankfull discharge, and meander geometry (Leopold *et al.*, 1964) as well as sediment size and transport and the type or style of river channel (Schumm, 1988). Thus, the morphology of a river system has been used to infer physical processes at a range of scales.

Fluvial geomorphology is the branch of science that attempts to find some systematic order in a suite of landforms and tries to understand the processes responsible for their development (Kellerhals et al., 1976), among other things. Given the wide range of river forms, this can be difficult, and it is an area of study that must move beyond the basic classification based on subjective, visual characteristics of a site or reach to one that is objective and includes interpretation of river system behavior (Thoms and Parsons, in review). Characterization and classification is an essential first step in defining and describing the physical properties of a river system and identifying the various components that influence their physical character. Once a river has been divided into appropriate homogenous areas, management objectives may be determined that are within the physical and biological limits of each area.

BACKGROUND PHILOSOPHIES AND APPROACHES

Many river science texts use the terms classification and characterization interchangeably. *Classification*, according to Bailey (1984), is the "general process of grouping entities by similarity," whereas *characterization* is merely describing the character of a single or group of entities (Brierley and Fryirs, 2005). *Taxonomy*, a somewhat related term, is an empirical, mainly objective procedure for allocating cases on the basis of their measured attributes, whereas *typology* is a conceptual process based upon a priori, subjective judgments of class definitions and boundaries. In terms of classifying rivers, Newson *et al.* (1998) suggested that these divisions – characterization, classification, taxonomy, and typology – are merely semantic. However, they do represent separate stages in a process of deriving a useful schema for the initial identification of distinct river types and then the allocation of new river types into appropriate classes of known morphologies and inferred behavior (Fig. 5.2). Whatever be the vocabulary, an enduring classification system should encompass broad spatial and temporal scales, integrate structural and functional characteristics under various disturbance regimes, convey information about underlying mechanisms controlling the riverine landscape, cost relatively little to implement, and reach a high level of consistent understanding among river scientists and managers (Naiman *et al.*, 1992). *Characterization* is the preferred term here and is used in the context of defining the process of identifying patches within the riverine landscape at various scales.

Traditionally, river characterization has relied upon the ordering of sets of observations or properties into meaningful groups based on their similarities or differences. Implicit in this exercise is the assumption that relatively distinct boundaries exist, and these may be identified by a discrete set of variables. Although river systems are continuously evolving and often display complexity, the grouping of a set of elements with a definable structure can aid in determining the influence of broader physical factors. This approach has also helped us understand why rivers have certain biological characteristics (Hawkes, 1975). Numerous approaches exist to assess the physical structure of river systems, some of which are hierarchical in nature (see Chapter 4), and have multiple levels that are useful for assessing river structure. A common practice is to divide rivers into homogeneous units (zones), within which the major controlling factors do not change appreciably, followed by further subdivision into smaller units in order to investigate the distribution of specific structures. This can be done normally with remotely sensed images, aerial photographs, and topographic maps, followed by detailed field studies.

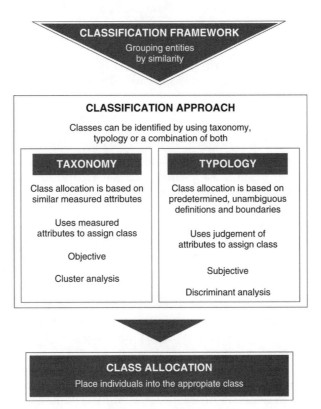

FIGURE 5.2 The classification process. After Newson *et al.* (1998).

No single classification scheme can satisfy all possible purposes or can encompass the multitude of river landforms. Regardless of the approach and methods employed, characterization schemes must be based on a sound conceptual *framework* underpinned by defensible scientific principles. A framework is neither a model nor a theory. Models describe how things work, and theories explain phenomena. In contrast, conceptual frameworks help to order phenomena and material, thereby revealing patterns (Rapport, 1985). Frameworks serve as scientific maps for new areas of endeavor; in this case, even tentative maps are useful (Pickett *et al.*, 1999) if only because their subsequent improvement provides some measure of progress in integrative thinking. In the context of river characterization, scientific principles must guide the process of identifying common river types and their distinguishing features as well as allocating river types to an existing characterization. Two important principles for the characterization of river systems are as follows:

Principle #1: Characterization of river systems must be undertaken at scales appropriate for the context in which they are to be used or the questions being asked. Riverine landscapes are the result of processes operating at multiple scales (Parsons and Thoms, 2007). Teasing apart regional and local effects requires appropriate stratification of sites along with the selection of variables at the correct scale for the study. Characterization should ideally be based on a holistic range of variables (*sensu* Biggs *et al.*, 1990) that are relevant to the physical character of the river system. Consequently, knowledge of the concepts of hierarchy theory is important here.

Principle #2: Identifying groups of interest must be based on the self-emergence of groups of similar character rather than imposing or inheriting groups from other studies or locations. Each scheme has its own inherent focus or context with which to study rivers and their character, and these will not be same for all studies. Characterization approaches must, therefore, evolve to being more objective rather than subjective.

Rivers have generally been characterized using one of two broad approaches. Multimetric methods involve the use of an array of variables or metrics at various scales that individually provide information on an array of physical attributes. When integrated, they are meant to provide an overall indication of the character of the river system. Multivariate methods also use a similar array of variables but employ statistical techniques, such as clustering analysis, discriminant function analysis (DFA), principal components analysis (PCA), and multidimensional scaling (MDS), to objectively determine the physical character of the river system. The advantages of multivariate methods are that they require no prior assumptions, according to Reynoldson *et al.* (1997), and they retain information that is lost in summary indices (Norris and Georges, 1993; Norris and Norris, 1995). Multivariate approaches are objective, thus providing higher levels of precision and accuracy than multimetric methods (Reynoldson *et al.*, 1997). They have been demonstrated to be an appropriate means for the characterization of sites (Gerritsen, 1995), although they are usually applied to smaller-scale structural features of river ecosystems (Chessman *et al.*, 1999). The disadvantages of multivariate approaches are that they are: (i) only inferential, unlike some statistical methods (Calow, 1992); (ii) sometimes considered overly complex (cf. Newson *et al.*, 1998), thereby requiring specialization by users; and (iii) difficult to convey to nonstatistically oriented river scientists and managers. Indeed, Gerritsen (1995) suggested that they can be daunting to all but experts in such methods because of the confusing multitude of techniques available and a lack of consensus on the best approaches.

The success of any river characterization requires a sound conceptual basis as to what is to be characterized and what are the goals of the process, along with rigorous design and statistical analysis. The scale used to underpin the characterization should be explicitly stated, as should the rationale for the choice of attributes. For choosing indicators and analyzing results, a multivariate approach is recommended over univariate techniques. Site selection for specific purposes should be based on a stratified random approach, with stratification established on a sound conceptual basis and power analysis used to determine the effectiveness of the sampling design. It is also important that the scientist initially specify the levels of statistical significance and power necessary to estimate the character of a river system at a specific scale (s) and identify the likely variance of condition and central tendency. The size of the smallest difference that needs to be detected should also be specified, and this degree of difference should be environmentally significant and scientifically attainable with realistically available resources.

DETERMINING THE CHARACTER OF RIVER NETWORKS: TOP-DOWN VS BOTTOM-UP APPROACHES

Top-Down Approaches

The majority of approaches to classify river channels tend to focus at the reach scale, as noted in Chapter 4. This weakens any attempt to classify or characterize entire river networks in a meaningful way that relates to influencing processes. Early classification schemes of river networks used broad criteria, such as the position of a reach within the catchment. However, there has been a recent tendency to employ more multimetric approaches in describing river networks. The concept of stream order proposed by Horton (1945) and later modified by Strahler (1957) is a common classification scheme. Stream ordering has become both a conceptual and an organizational tool in many areas of river science. However, comparisons between channel networks can prove misleading for a number of reasons. An order assigned to a channel segment depends on the criteria used to determine where first-order channels begin within a network, and no accepted standard exists for this task. Furthermore, the representation of a stream network in a catchment varies depending on the scale of maps used to derive the stream network. In the United Kingdom, Gardiner (1995) recommended using 1:25,000 maps, whereas other studies have employed 1:100,000 scale maps (Patil, 2002). Moreover, in a recent study to delineate river networks in the Maloti-Drakensberg region of South Africa, Dollar *et al.* (2007) relied on a collective of 1:100,000 and 1:150,000 scale maps. Although interesting differences were detected between subcatchments, these apparent differences were confounded by varying map scales. As Montgomery and Buffington (1998) noted, channel networks defined from the blue lines on a map along with information on the curvature of the contours or a critical gradient or drainage area can differ substantially from the network identified in the field. In addition to these scale problems, the main criticism of the concept of stream ordering is the absence of an inherent association between a stream order number and channel morphology or the operating fluvial processes. Stream ordering provides only an indication of relative channel size and position within the channel network.

Many channel classifications schemes are based upon general differences in channel patterns and processes (cf. Chapter 4), and these have been used to describe river networks. Leopold and Wolman (1957), for example, differentiated between straight, meandering, and braided channel patterns on empirical relationships between slope and discharge, while Schumm (1977b) classified alluvial channels based on dominant modes of sediment transport (i.e., suspended, mixed, or bed load transport). The latter has been used as a basis to define three broad functional regions or geomorphic zones within catchments (cf. Fig. 5.3). Headwater regions are recognized as the primary sediment supply area. In these regions, the controlling processes are weathering and the downslope movement of this weathered material. The presence of narrow valleys and the lack of floodplains in headwater regions provide a high degree of connectivity between hillslopes and river channels, hence their recognition as a net supplier of

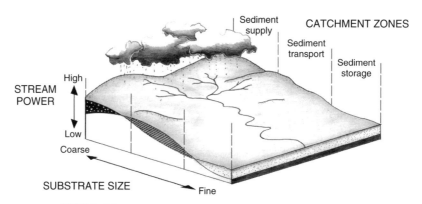

FIGURE 5.3   Catchment process zones based on Schumm (1977a).

water and sediment within a river network. By comparison, in the lowland regions, where river slopes and associated stream energy decrease dramatically, sediments are generally deposited, forming large floodplains. These wide floodscapes are often dissected by various river channel patterns, including meandering, anabranching, and distributary channels. This region, of a catchment, is often referred to as a depositional or storage area. The boundary between the headwater and lowland regions is where river slopes decrease, valley floor widens, and the dynamic nature of the river increases. This has been designated as a transfer area where sediments can be both highly mobile and temporarily stored within and adjacent to the river channel. Associated with these changes in the flow and sediment regime throughout a catchment are changes in river morphology and their behavior.

Classifications generally use similarities of physical form and process or function to impose order on a continuum of natural stream types or morphologies. Approaches employed to classify rivers serve various purposes, and no single classification approach or scheme can satisfy all possible purposes or can encompass all channel types. Indeed, few provide a comprehensive assessment of river channels at multiple scales. Exceptions include that based on the fluvial hydrosystem concept, as developed in Chapter 4, and the hierarchical channel classification approach of Montgomery and Buffington (1998), which attempts to address different factors influencing channel properties at multiple scales. In the latter, the hierarchy of spatial scales reflecting differences in processes and controls on channel morphology are used, and these included the geomorphic province, catchment, valley segment, channel reach, and channel unit (Table 5.1). *Geomorphic provinces* are broad scale regions with similar landforms, which reflect compatible hydrologic, erosional, and tectonic processes. *Catchments* within the same province tend to share similar relief, climate, and lithologic assemblages. Catchments are spatial, bounded entities that define the river network and therefore the routing of water and sediment across the landscape. Nested within the catchment are *valley segments*, which are portions of the catchment network exhibiting comparable valley morphologies and governing geomorphic processes. *Channel reaches* are nested within valleys segments, and they exhibit similar bedforms over stretches of stream that are many channel widths in length. At the smallest scale are *channel units*, which are morphologically distinct areas extending up to several channel widths in length and represent the physical building blocks of a reach. Typical channel units include various types of pools, bars, and shallows, that is, riffles, rapids, and cascades (Bisson *et al.*, 1982). Distinctions among channel units are usually made on topographic form, organization, local slope, flow depth and velocity, and sediment texture.

This approach can be and has been used to classify river networks, albeit for mountain streams only. Implicit in this exercise is the assumption that relatively distinct boundaries exist and these may be identified by a discrete set of variables. At the network scale, Montgomery and

TABLE 5.1   Hierarchical Levels of Organization for Use in
Channel Classification and Their Associated Spatial Scales

Classification level	Spatial scale
Geomorphic province	$1000 \, km^2$
Watershed	$50–500 \, km^2$
Valley segment	$10^2–10^4 \, m$
Colluvial valleys	
Bedrock valleys	
Alluvial valleys	
Channel reaches	$10^1–10^3 \, m$
Colluvial reaches	
Bedrock reaches	
Free-formed alluvial reaches	
Cascade reaches	
Step-pool reaches	
Plane-bed reaches	
Pool–riffle reaches	
Dune–ripple reaches	
Forced alluvial reaches	
Forced pool–riffle reaches	
Channel units	$10^0–10^1 \, m$
Pools	
Bars	
Shallows	

After Montgomery and Buffington (1998).

Buffington (1998) relied on position within the network, differences in the ratio of transport capacity to sediment supply, and the identification of source (colluvial segment), transport (bedrock segment), and response zones (alluvial valley segment) (Fig. 5.4) (Bisson and Montgomery, 1996). These valley segments represent relatively homogeneous lengths of the river network within which the major controlling factors (the flow and sediment regimes) do not

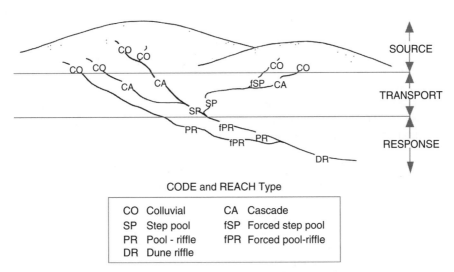

CODE and REACH Type

CO	Colluvial	CA	Cascade
SP	Step pool	fSP	Forced step pool
PR	Pool - riffle	fPR	Forced pool-riffle
DR	Dune riffle		

FIGURE 5.4   River reaches based on process domains of Bisson and Montgomery (1999).

change appreciably. These different valley segments can be identified from aerial photographs and maps, but field verification is required. However, in their study of the South Fork River, Washington, USA, the different valley segments were delineated by local slopes (S) derived from digital elevation data, i.e., areas of source ($S > 0.30$), transport ($0.03 < S < 0.30$), and response ($S \sim 0.03$). Having established the broader-scale valley segments, individual channel types are then identified. The most common channel types identified by this approach (e.g., Cupp, 1989; Bisson and Montgomery, 1996; Montgomery and Buffington, 1998) include the following:

- *Colluvial channels* are small channels surrounded by colluvium and are generally located in the upper reaches of channel networks.
- *Cascade channels* are the steepest of the alluvial channels and are characterized by large sediment textures that form the primary roughness elements and impose a strong three-dimensional structure to the flow.
- *Step-pool channels* are fully occupied with accumulations of coarse sediment and form a sequence of steps, typically one to four channel widths apart. Both "free" and "forced" step-pool channels can be identified, depending on whether they are comprised of alluvial sediment or immovable obstructions from the majority of the steps.
- *Plane-bed channels* have no well-defined bedforms but display long, and commonly channel-wide, reaches of uniform riffles or glides.
- *Pool–riffle channels* are the most common form of lowland channels, with distinctive pool–riffle sequences.
- *Dune–ripple channels* represent the classic lowland sand-bedded channels and are typically found in large rivers, where the character of the predominant bedform will change in response to increasing discharge from plane bed at low flows to ripples, sand waves, dunes, high-energy plane bed, and antidunes at highest flows.

Field observations in the mountain drainage basins of the Pacific Northwest of the United States suggest channel slope to be an important discriminator of these channel types. Pool–riffle channels occur at gradients <0.02, plane-bed channels between 0.01 and 0.03, step-pool channels between 0.03 and 0.08, and cascade channels between 0.08 and 0.30. In contrast, colluvial channels are present at gradients >0.20, and bedrock channels have unusually steep slopes for their drainage area in the mountainous drainage basins of this region (Bisson and Montgomery, 1996).

Valley segment classifications based on valley cross-sectional shape, valley bottom gradient, channel pattern, and channel confinement have been used in many studies to stratify channels within a river network (e.g., Paustian *et al.*, 1984; Frissell et al., 1986; Cupp, 1989). However, these approaches have relied on data and information gathered from intensive field surveys, which may require considerable time for collation. Recent advances in the development of relatively rapid objective river classification schemes for river networks have occurred, and this has paralleled advances in the availability of high-quality digital elevation data, computer processing abilities, and Geographic Information Systems (GIS) tools. The approach developed by Thoms *et al.* (unpublished data) is an example of these and one that uses readily available data sets (DEMs and streamline network models) along with information on catchment geology and climate to derive physical classification for rivers and floodplains at the valley scale – a scale that is appropriate for classifying entire river networks. It is underpinned by hierarchy theory and, therefore, focuses only on those variables relevant to identifying valley-scale river zones within a network. The philosophy and approach is outlined below.

Many interrelated geomorphological factors operate within the riverine landscape. These factors sit within a hierarchy of influence, where larger-scale factors set the conditions within which smaller-scale factors operate (Schumm and Lichty, 1965). According to the riverine landscape hierarchy of Schumm and Lichty (1965), climate and geology control the physiography of the catchment and the types of vegetation and soils that are present, and they indirectly influence how humans use the catchment surface (see Table 3.2). Catchment physiography, vegetation, soils, and landuse then control the important channel-forming factors of flow and sediment supply and transport as well as the slope of the valley, bank characteristics, and sediment caliber. These channel-forming factors subsequently control channel geometry and flow factors that occur at the bottom of the cascade, such as channel slope, water velocity, river planform, cross-sectional channel dimensions, and the arrangement of bedforms in the channel (see Table 3.2). Thus, in a fluvial system, geomorphological factors operating at one level of the hierarchy constrain the formation of factors at successively lower levels (Schumm, 1991).

Overlain across this cascade of interrelated geomorphological factors is a temporal context that determines the strength of constraint between successive hierarchical levels (Schumm and Lichty, 1965). The relationship between factors located at different levels of the hierarchy is a function of the persistence of each factor over time. In general, the strength of influence between factors sitting at the top and bottom of the geomorphological hierarchy decreases at longer time spans (see Table 3.2). The independent catchment factors, climate and geology, perch at the top of the hierarchy because over a geologic time span of millions of years, they constrain the formation of factors such as vegetation, relief, palaeohydrology, and valley dimension. However, at this time span, factors that sit at the intermediate and lower levels within the cascade, such as sediment discharge, water discharge, channel morphology, and flow character, are not influenced by any of the higher-level factors (see Table 3.2). Over a time span of thousands of years, channel morphology becomes the only factor that is influenced by factors operating at higher levels of the cascade, and lower-level factors such as sediment discharge, water discharge, and flow character are unresponsive. This pattern of constraint continues until at a time span of 1 year, sediment discharge, water discharge, and flow character are the only factors that are influenced by higher-level factors on a yearly basis (see Table 3.2). Thus, interrelationships between any of the geomorphological factors in a river system are directly linked to the evolution and behavior of the system at different temporal scales.

This concept of river classification is based on this cascade of geomorphic interrelationships that operate across various scales. Although most geomorphic characterization systems for rivers are inherently hierarchical (e.g., Frissell *et al.*, 1986; Rosgen, 1994; Brierley and Fryirs, 2005), they typically fail to use the hierarchical arrangement for detailed river classification exercises (but see exceptions in Montgomery and Buffington, 1998; Snelder and Biggs, 2002). That is, classification variables are often included from many levels within the hierarchy rather than from a narrow band of levels that bracket the scale of interest. This may result in erroneous classifications that are based on properties that are insignificant at the scale of interest. Consequently, only characterizations with an explicit focus on scale are likely to accurately capture the true complexity of river systems and will reflect only those processes and/or forms that are most relevant to the scale of interest. It follows then that in applying a hierarchical approach to classification for the entire riverine ecosystem, river types must be derived on the basis of catchment, valley, and channel variables – that is, those variables operating at a scale commensurate with the scale of interest ... the river network.

Recent river characterization approaches have tended to be quantitative in approach, employ data gathered from multiple sources, and use advances in modern technology. As an example, the approach by Thoms *et al.* (unpublished data) is outlined here. The main data sets required for the approach include high-quality DEMs (e.g., 30 m or better) of the catchment and accompanying streamlines. These data along with geological and climatic information are used to generate a data matrix required for the classification of the river network. Fifteen variables spanning three scales of the riverine landscape hierarchy are determined at specific sites throughout the river network (Table 5.2). Sites are located at known intervals along the various streamlines that comprise the river network. The spacing of the sites can vary depending on computer processing capacity and/or the quality of the data. In the catchments classified by Thoms *et al.* (unpublished data), a point creation tool was developed and employed to generate

TABLE 5.2 Independent and Dependent Variables Used to Derive Different FPZs

Variable	Variable type	Description
Geology	Independent	Bedrock, alluvium or nonalluvium character
Mean annual rainfall	Independent	Average annual rainfall (mm)
Elevation	Independent	Height above the mean sea level
Valley width	Independent	Width of the valley (m)
Valley floor width	Independent	Width of the valley floor (m)
Valley side slope (includes left and right hillslopes)	Independent	Slope of the adjacent hillslopes
Down-valley slope	Independent	Slope of the valley in a downstream direction
Ratio of valley to valley floor width	Independent	Ratio of the valley width to the valley floor width
Wavelength of the channel belt	Dependent	Meander wavelength of the channel belt (m)
Sinuosity of the channel belt	Dependent	Sinuosity of the channel belt
Width of the river channel belt	Dependent	Width of the river channel belt (m)
Sinuosity of the river channel	Dependent	Sinuosity of the river channel
Number of channels	Dependent	Number of channels
Channel planform	Dependent	Anastomosing or meandering

From Thoms *et al.* (unpublished data).

the series of sites at 10-km intervals along the streamline layer. These sites become the locus for data extraction for the catchment, valley, and channel variables.

For the catchment scale variables, the dominant geology related to each site can be either a categorical variable, with the dominant geology being classed as either alluvial, nonalluvial, or bedrock, or the actual geological composition of a site determined. Effective rainfall is considered to be the main climate variable determined at each site, and this can be obtained from long-term rainfall data available for most catchments. The seven valley-scale variables used for the classification included the following: elevation, valley width, the width of the valley floor trough, ratio of the valley width to the valley floor trough width, valley slopes, and down-valley slope. Site elevations are determined from the catchment DEM, while valley width is measured between the highest point on each of the valley divides, as obtained from the DEM. The valley floor trough can be identified as the intersection between the adjacent hillslope and the valley bottom. Finally, the down-valley slope and sinuosity are computed for predetermined distances between successive sites. These valley data can be verified by overlaying the valley information on SPOT imagery to check for consistency. In addition to the catchment and valley variables, six channel variables are used (see Table 5.2), these being: the following: channel sinuosity, channel planform class, the number of channels present at a site, channel belt sinuosity, channel belt width, and channel belt wavelength.

Site data collected for the river network can be analyzed in several ways. Standard multivariate analyses of the entire data set can be undertaken to reveal groups of sites with a similar character. For example, Thoms *et al.* (unpublished data) classified data for the Ovens River in southeastern Australia, using a fusion strategy recommended by Belbin and McDonald (1993) termed flexible unweighted pair-groups with arithmetic averages (UPGMA). For this routine, the Gower association measure is recommended because it is range-standardized and is applicable for nonbiological data (Belbin, 1993). Groups of sites with similar physical character can be selected by viewing the resultant dendrogram representation of the classification. Once identified from the dendrogram, groups of similar sites are then arrayed onto the streamlines or the river network, thus providing the spatial distribution of common river types within the river network (cf. Fig. 5.5). These groups of sites identified in the classification analysis represent *FPZs* – lengths of river with similar valley–floodplain settings and river morphologies – and are inferred to be influenced by similar geomorphic processes.

This type of analysis is, however, dependent upon the selection of a threshold dissimilarity value. It ultimately determines the number of groups or FPZs that are identified from the

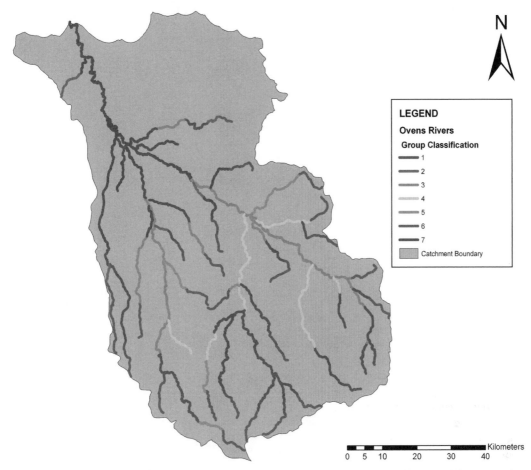

N

FIGURE 5.5   River zones within Ovens catchment, Australia. These river zones were determined from a top-down approach with methods described in the text. (See color plate 5).

dendrogram. This can be done objectively in several ways and thus overcomes the often subjective nature of determining patterns from such an analysis. An a priori dissimilarity value can be used, which Quinn and Keough (2002) suggested should correspond to a level of 80% similarity. However, reasons for why certain priori levels were selected are never explained in any detail in the literature and often attract a level of criticism (Newson *et al.*, 1998). Dendrograms rarely display a linear increase in the number of groups; more commonly the number of groups increases exponentially with increasing similarity. An alternate approach to the selection of a threshold similarity value considers the relationship between the number of groups in the classification and the dissimilarity value. These relationships generally have an inflexion point where the increase in the number of groups is greater beyond a certain similarity value. Figure 5.6 provides several examples of this. This approach has been successfully applied in several different environmental applications including sediment textural analysis (Forrest and Clark, 1989) and catchment hydrology (Thoms and Parsons, 2003).

Once the number of groups has been determined, further data analysis can be undertaken to assess their validity. Groups of similar sites can also be ordinated using semistrong-hybrid multi-dimensional scaling (Belbin, 1993), then tested whether the ordination solution (as determined by the stress level) occurred by chance alone. Relationships between the physical variables used in the

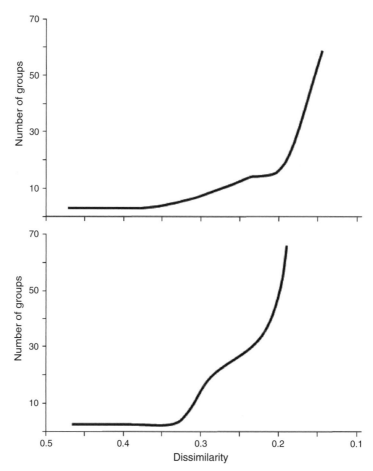

FIGURE 5.6 Relationships between dissimilarity value and the number of cluster groups within a dendrogram. Inflection points are recognized as the basis for separating groups of similar FPZs.

analysis and groups of sites in ordination space can also be investigated using principal axis correlation (PCC) (Belbin, 1993). Principal axis correlation generates a correlation value (R^2) for each physical attribute used in the analysis, with high values being indicative of a strong association between a physical variable and the position of stations in ordination space. A Monte Carlo permutation test should also be applied to test the significance of the correlation values. Only the significant variables with an R^2 above the 75th percentile are recommended.

Bottom-Up Approaches

Quantitative methods have been employed for data collected at smaller scales: the functional unit and mesohabitat scales. These approaches can be referred to as being *bottom-up-driven*, where sections of the river network are grouped together based on similarities of within-river channel features rather than on similarities of the valley and floodplain settings and channel morphologies. The river habitat survey (RHS), for example, was designed to assess the quality and variability of physical habitat of stream and rivers throughout Britain. Full details of survey, which assessed over 5000 sites, are provided in Raven *et al.* (1998a). The broad categories of geomorphological information in the RHS included data on channel dimensions, bed and bank material, bank and channel features, and basic data on channel planform. These local scale data were employed by Newson *et al.* (1998) for objective classification of British rivers. The core analytical procedure employed, in this geomorphological classification, was a statistical classification using TWINSPAN (two-way indicator species analysis; cf. Hill, 1979) accompanied by redundancy and discriminant analyses. Essentially TWINSPAN yields a cluster dendrogram, similar to that noted in the previous section along with an analysis of those variables that are associated with the divide between various groups. In this study, the analysis was considered to be useful as it was able to delineate British rivers on the basis of substrate size and stream power. Application of these results by Newson *et al.* (1998) allowed the development of a stability index that discriminated between river reaches that exhibited "stable" geomorphological properties and those that showed evidence of dynamic adjustment. A total of 16 stability classes contained in four broad groups were generated, the spatial distribution of which is displayed in Fig. 5.7.

Approaches and methods of classification put forward thus far are dominated by a philosophy of identifying commonalities (similarities) in the physical attributes that reflect the character of the riverine landscape at a particular scale. However, recognition of patches in a riverine landscape infers the presence of edges or boundaries between patches, such as FPZs or river reaches. A landscape boundary can be defined as a zone of transition between adjacent ecosystems or in this case, lengths of river that have a set of physical, chemical, and even biological characteristics uniquely defined by space and timescales. Boundaries are locations where the rates or magnitudes of transfers change abruptly in relation to those within respective patches and have been recognized operationally by spatial discontinuities or areas of high variance in physical and/or biological character (Holland, 1988; Gosz, 1991). Boundary formation can result from the same processes that produce patches, namely geomorphology, geology, resource distribution, and the disturbance regime. A patchy physical environment, such as a mosaic of river zones, will produce numerous boundaries separating relatively homogenous patches. Likewise, patchiness in resource distribution, for example, nutrients in a landscape or carbon within riverscapes, will also create boundaries. These physical and chemical boundaries usually manifest themselves in distinct energy, vegetation, and biotic distribution patterns. Disturbance, such as flooding, plays a large role in boundary formation as it has the potential to dramatically alter the physical, chemical, and biological environment (White and Harrod, 1997). Landscape boundaries are often patchy in nature, with small patches of one patch type surrounded by the second patch type on both or one side of the border (Forman, 1997). Thus identifying boundaries or discontinuities in environmental character can also lead to the identification of individual FPZs within river networks.

Locations of high variance or discontinuities can be determined by several approaches, and these have been well-developed by terrestrial ecologists (e.g., Ludwig and Cornelis, 1987; Forman, 1997). When morphological data are available for entire lengths of river channel, locations of high variance or channel boundaries can be determined using a moving window technique (Ludwig and Cornelis, 1987). The size of the window is important, and it refers to the number of samples or locations along a river network for which a change in variance is evaluated. Essentially, the mean for each half-window is subtracted from the mean of adjacent half-windows to determine the variance between the two half-windows in terms of Euclidean distance along a sequence. The window is moved one sample point or location along a river network and the process is repeated until the entire river network has been sampled. This type of edge or boundary detection has been used with great success in detecting vegetation boundaries (REFS), and Ludwig and Cornelis (1987) recommended experimenting with various half-window sizes.

Moving window analysis was applied by Thoms (unpublished data) to determine boundaries between river zones for a 2500-km section of the Barwon–Darling River in southeastern Australia. The Barwon–Darling River is a dryland system with significant areas contributing no runoff (Thoms *et al.*, 2004). It is a suspended load river (cf. Schumm, 1977a), with a relatively

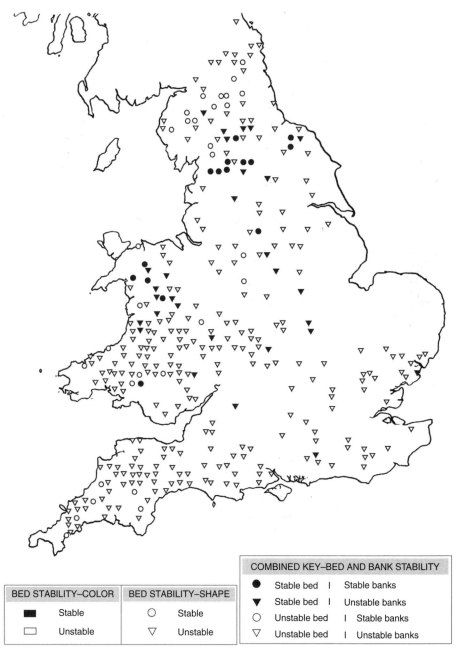

BED STABILITY–COLOR		BED STABILITY–SHAPE	
■	Stable	○	Stable
▢	Unstable	▽	Unstable

COMBINED KEY–BED AND BANK STABILITY			
●	Stable bed	|	Stable banks
▼	Stable bed	|	Unstable banks
○	Unstable bed	|	Stable banks
▽	Unstable bed	|	Unstable banks

FIGURE 5.7   Identified river reaches in England and Wales based on the RHS. These reaches were identified using site data. From Newson *et al.* (1998).

deep (average bankfull depths of 6–20 m) and narrow (average bankfull widths of 50–90 m) channel that is highly sinuous. Although there are no major inflows along large sections of the Barwon–Darling, significant changes in channel morphology still occur. To quantitatively determine the distribution of FPZs along the river, this study used 850 bankfull cross sections spaced at somewhat regular intervals along the river. For each cross section, 12 morphological variables describing the size, shape, and complexity of the bankfull channel were calculated

TABLE 5.3 Morphological Variables Used to Describe the Bankfull Cross Section

Cross–sectional character	Variable description
Cross–sectional size	Bankfull width
	Bankfull depth
	Cross–sectional area
	Wetted perimeter
Cross–sectional shape	Width/depth ratio
	Shape index
	Hydraulic radius
Cross-sectional complexity	Irregularity
	Complexity index
	Bench index
	Number of bench surfaces
	Flow conveyance factor

Descriptions of the variables and the formulae for calculating them can be found in Thoms (unpublished data).

(Table 5.3). These 12 morphological variables provided the basic data to calculate the Euclidean distance measure (Equation 5.1) for the channel morphology.

$$\text{XsecED} = 1 - \sqrt{\frac{[(1-A)^2+(1-W)^2+(1-D)^2+(1-Wp)^2+(1-Hr)^2+(1-Rcs)^2+(1-NoB)^2+(1-Cx)^2+(1-Cxbd)^2+(1-Cxbk)^2+(1-Cy)^2+(1-Z)^2]}{\sqrt{n}}}$$

(5.1)

where:

XsecED = Euclidean distance for the bankfull cross section
A = Area of the bankfull cross section
W = Width of the bankfull cross section
D = Depth of the bankfull cross section
WD = Width to depth ratio of the bankfull cross section
Wp = Wetted perimeter of the bankfull cross section
Hr = Hydraulic radius of the bankfull cross section
Rcs = Shape of the bankfull cross section
NoB= Number of benches with the bankfull cross section
Cx = Complexity of the bankfull cross section
Cxbd = Complexity of the bed of the bankfull cross section
Cxbk = Complexity of the river banks of the bankfull cross section
Z = Conveyance factor of the bankfull cross section
n = Number of components

The resultant moving window of the Euclidean distance for the Barwon–Darling displayed consistently low variance values for large sections of the river (Fig. 5.8), and these indicated areas of relative homogeneity between cross section or river zones with similar cross-sectional morphology, that is, river patches. Peaks in Euclidean distance, however, indicated areas of high variation and, therefore, boundaries between river zones. Several small peaks in Euclidean distance were noted, but these represented small-scale background variance in the data as per that noted by Ludwig and Cornelis (1987).

Regardless of whether top-down or bottom-up approaches are used, river classification schemes have extensively exploited the physical features of the riverine landscape to identify commonalities at various scales. These commonalities are used to infer the influence of similar

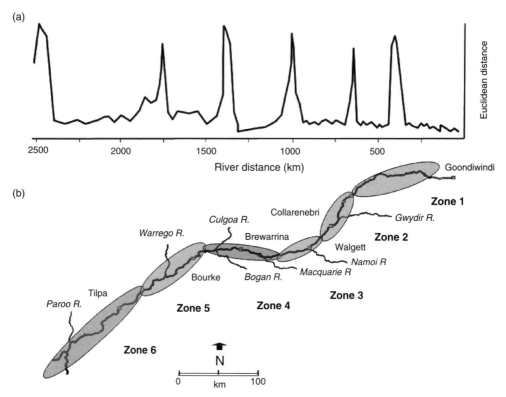

FIGURE 5.8   The river zones of the Barwon–Darling River, Australia: (a) moving window analysis of bankfull cross-sectional variables and (b) location of the river zones. Note that the location of each FPZ does not correspond with tributary junctions. Redrawn from Thoms (unpublished data).

physical processes, this being based on the premise that rivers are process-response systems. Studies have also used process variables, principally stream flow, to describe and classify the physical environment of rivers. Because of the ways flow can be characterized, analyzing stream flow variables using a multivariate approach is an effective means to determine similarities among river networks (Pegg and Pierce, 2002). This style of approach to river classification has potential, especially when large spatial and temporal data sets are available. Using a multivariate approach, Pegg and Pierce (2002) were able to group flow-gauging sites into six hydrologically similar units in the Missouri and lower Yellowstone Rivers. This study was limited to 15 flow-gauging stations and the suite of hydrological variables prescribed in the Indicators of Hydrologic Alteration (Richter *et al.*, 1996). This approach has been refined and extended by others, notably Thoms and Parsons (2003), Poff *et al.* (2006), and Leigh and Sheldon (2008).

Flows are monitored, usually on a daily basis, in many river basins, and many countries have multiple flow stations along individual rivers. These data can be used to investigate and determine river zones. The example provided here from the Condamine Balonne River, Australia, modeled daily flow (Fig. 5.9) from 43 flow stations for the period 1922–1995. Full details of this study are given in Thoms and Parsons (2003). A full description of the hydrological model and its reliability is detailed in Thoms and Parsons (2002) and Thoms (2003).

A full set of 340 flow variables were calculated for each of the 43 flow stations. These flow variables fell within various categories encompassing magnitude, frequency, duration, timing, and rate of change aspects of the flow regime (Table 5.4). Each of the categories can be assigned to the regime, history, or pulse scale according to the time period over which they influence

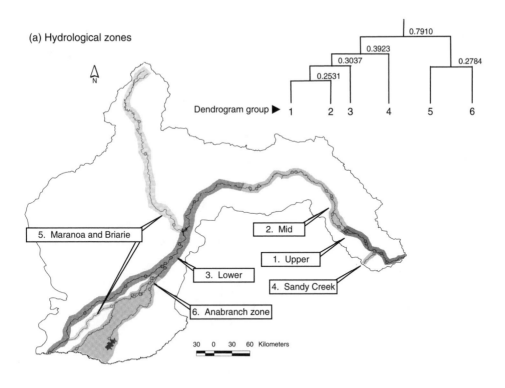

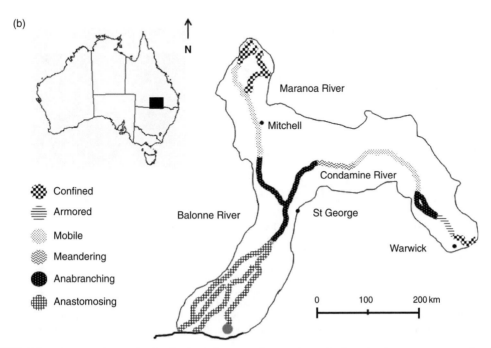

FIGURE 5.9   River zonation in the Condamine Balonne catchment, Australia. Functional process zones are based on (a) hydrological character and (b) river channel morphology. Modified from Thoms and Parsons (2002).

TABLE 5.4 Flow Variables Used by Thoms and Parsons (2003) in Their Flow Classification Procedure

Variable type, Variable category	Scale	Acronym (and number of variables in each category)	Number of variables included in the pre- and current scenario data sets
Daily flow summary			
Long-term values	Regime	LT (1–18)	15
High-flow spell analysis			
Number of "above-threshold" flows	History	HSN (1–22)	17
Peak magnitude of "above-threshold" flows	Pulse	HSP (1–22)	17
Duration of "above-threshold" flows	Pulse	HSD (1–22)	17
Seasonal variation of "above-threshold" flows	History	HSSV (1–36)	30
Low-flow spell analysis			
Number of "below-threshold" flows	History	LSN (1–14)	8
Trough magnitude of "below-threshold" flows	Pulse	LST (1–14)	2
Duration of "below-threshold" flows	Pulse	LSD (1–14)	8
Seasonal variation of "below-threshold" flows	History	LSSV (1–36)	18
Moving average			
Maximum annual moving average	Regime	AMAX (1–12)	9
Minimum annual moving average	Regime	AMIN (1–12)	4
cessation of flow analysis			
Zero flows	History	ZF (1–6)	5
Number of rises and falls of the hydrograph	History	RFN (1–8)	8
Duration of rises and falls of the hydrograph	Pulse	RFD (1–8)	8
Magnitude of daily change in flow	Pulse	RFM (1–32)	26
Monthly flow analysis			
Monthly flows	Regime	MF (1–50)	26
Intermonthly variability	History	MFMV (1–4)	2
Interannual, monthly variability	Regime	MFAV (1–4)	4
Additional variables			
Number of days between spells of zero flow	History	NDAY0 (1)	1
Maximum number of days between spells of 1/2 mean daily flow	History	NDAY5 (1)	1
Maximum number of days between spells of 1/3 mean daily flow	History	NDAY3 (1)	1
Maximum number of days between spells of 1/9 mean daily flow	History	NDAY9 (1)	1
Mean annual flow	Regime	MAFL (1)	1
CV mean annual flow	Regime	CVMAF (1)	1

Seven main types of variables were included, containing various categories of variables. Acronyms correspond to each category, but are also numbered sequentially within a category. For example, within the "long-term values" category (LT), there were 18 individual flow variables. A full list of 340 flow variables can be obtained from the authors.

hydrological character (Thoms and Parsons, 2003). For example, *maximum annual moving average* includes regime-scale variables (see Table 5.4) because it represents macroscale hydrological influences occurring over hundreds of years. Similarly, *seasonal variation of below-threshold flows* is a history-scale variable (see Table 5.4) because this metric describes flow characters over tens of years, and *magnitude of daily change in flow* encompasses pulse-scale variables (see Table 5.4) of hydrological change over daily periods. The number of hydrological variables that can be used in this approach will depend on the quality of data available and the nature of the project. Although Thoms and Parsons (2003) used all 340 variables, other studies (Maddock *et al.*, 2004; Sheldon and Thoms, 2006a) have used many fewer. We recommend using the scales rather than types of flow variables as the basis for examining the temporal dimension of hydrological character in river systems and for considering the spatial organization of flow stations that may have similar hydrological character.

Some of the flow stations contained missing data, arising from zero-divide errors in the calculation of some flow variables, and these must be removed from any subsequent analyses. Multivariate statistical techniques generally require a complete data set that is free of missing values; therefore, any flow variable containing missing data values for one or more of the 43 flow stations was deleted from the data set. Flow variables that are invariant (i.e., they contained the same value across all of the flow stations) must also be removed from the data. For the Condamine Balonne study, the final data set for the multivariate analysis contained 230 flow variables.

The multivariate analyses for the Condamine Balonne study were performed using the PATN analysis package (Belbin, 1993). However, other commercial packages for analysis of this type are available, each with its own strengths and weaknesses. In this example, data for the 43 flow stations were classified using the flexible UPGMA fusion strategy, as recommended by Belbin and McDonald (1993) for large data sets. The Gower association measure was used in all classifications, because this measure is a range-standardized measure that has been recommended for nonbiological data (Belbin, 1993). There are other associated measures, but many do not provide standardization within and between variables. Groups of flow stations with similar hydrological character were selected by viewing a dendrogram representation of the classification. The selection of a standard dissimilarity threshold level is common in many studies (cf. Quinn and Keough, 2002), but in many cases, this can lead to the identification of groups of sites, which has no practical meaning. The construction of a simple relationship between the dissimilarity level and the cumulative number of groups can assist with the selection of groups from the dendrogram. In particular, any point of inflexion should be noted. Once the dendrogram groups are selected, these can be arrayed onto a map of your river system to delineate the position of the various flow stations with similar hydrological character. These groups of flow stations with similar hydrological character equate to hydrological zones.

In the Condamine Balonne study, the flow stations were then ordinated using semistrong-hybrid multidimensional scaling, and a Monte Carlo permutation test (Belbin, 1993) was performed to determine whether the ordination solution (as determined by the stress level) occurred by chance alone. Flow stations were arrayed in ordination space according to the dendrogram groups identified in the classification analysis, which in turn equate to hydrological zones. The ordination was performed in three dimensions, and as such, there were three possible axis comparisons. Biplots of each combination were constructed, and the axes representing the best separation of dendrogram groups in ordination space were selected for presentation. This was done to assess if there were any difference between the different hydrological zones.

To determine the relationship between the flow variables used and groups of flow stations in ordination space, a PCC was used (Belbin, 1993). Principal axis correlation generates a correlation value (R^2) for each attribute, with high values being indicative of a strong association between a flow variable and the position of the groups of flow stations in ordination space. A MonteCarlo permutation test (Belbin, 1993) can also be performed to test the significance of the correlation values. Only significant variables (R^2 greater than the 80th percentile) should be used (Thoms and Parsons, 2002). Some of the vectors will be from the same category and, thus, highly correlated. However, in the Condamine Balonne study, no attempt was made to identify and remove correlated flow variables because it was desirable to examine if a dominant scale of variable was strongly associated with groups of nodes in ordination space. Association between each vector and a group of flow stations was assigned visually, and vectors were subsequently tallied according to the regime, history, and pulse scales. This allows for the identification of temporal scales of hydrological influence for the different hydrological zones.

Classification of 43 flow stations in the Condamine Balonne revealed six groups with similar hydrological character (see Fig. 5.9a). Group 1 flow stations were located in the upper section of the river, while Group 2 flow stations were situated further downstream in the midsection of the river. Group 3 flow stations were located in the lower section of the river, and this included the main Condamine Balonne River channel along with the Culgoa and Balonne-Minor River anabranch sections. Group 4 contained only one flow station: Sandy Creek, which is a tributary within the upper section of the catchment. Group 5 contained flows stations from the Maranoa River but also included the only flow station present in Briarie Creek. Group 6 flow stations were all located in the anabranching lower section of the Condamine Balonne River. Thus, the six groups of flow stations corresponded to distinct zones of hydrological character within the Condamine Balonne catchment.

These hydrological zones corresponded closely with an independent river classification of the Condamine Balonne undertaken by Thoms and Sheldon (2002). The upper hydrological zone (see Fig. 5.9b) corresponded with the constrained upland and armored geomorphological zones. In the remainder of the main channel, the mid, lower, and anabranch hydrological zones (see Fig. 5.9a) corresponded with the mobile, meandering, and anabranch geomorphological zones, respectively (see Fig. 5.9b). Thus, there is an association between the independently derived hydrological and geomorphological zones of the Condamine Balonne River.

The hydrological zones of the Condamine Balonne form clear groups in ordination space (see Fig. 5.9b). Fifty flow variables were subsequently identified as being highly correlated with each ordination space and were associated with individual hydrological zones (see Fig. 5.9a). In the upper zone, history and pulse-scale flow variables were prominent, and in the anabranch zone, regime, history, and pulse-scale variables were more salient (see Fig. 5.9a). In addition to these minor scale-related trends, specific types of variables were associated with hydrological zones. In the upper zone, variables representing the magnitude and seasonal variation of high-flow spells were important (see Fig. 5.9a). Variables representing the magnitude and duration of high-flow spells, the duration and seasonal variation of low-flow spells, and the number and magnitude of rises and falls were important in the mid and lower hydrological zones. In the anabranch zone, the duration and number of high-flow spells was important.

Comparing Top-Down vs Bottom-Up Approaches: An Example

Brungle Creek is a right-bank tributary of the Tumut River, which in turn is a dominant headwater catchment in terms of water yield to the Murrumbidgee system in the Murray–Darling Basin. With a catchment area of $1364.5\,km^2$ and a mean annual rainfall of 1075 mm, Brungle Creek is a typical Australian headwater system with elevations greater than 1000 m above mean sea level. The creek flows through several different valley settings before joining the Tumut River. The upper reaches are constrained by local faults, while further downstream variable influences of different geologies produce a series of alternating unconstrained and constrained valleys. At 60 sites along the creek, bankfull cross sections were surveyed, 16 variables (as described previously and listed in Table 5.4) were calculated, and a moving window analysis was applied. The resultant profile of the Euclidean distance had three major peaks (Fig. 5.10), and these were interpreted to separate four different FPZs within Brungle

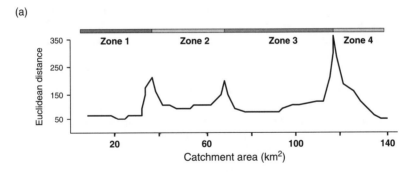

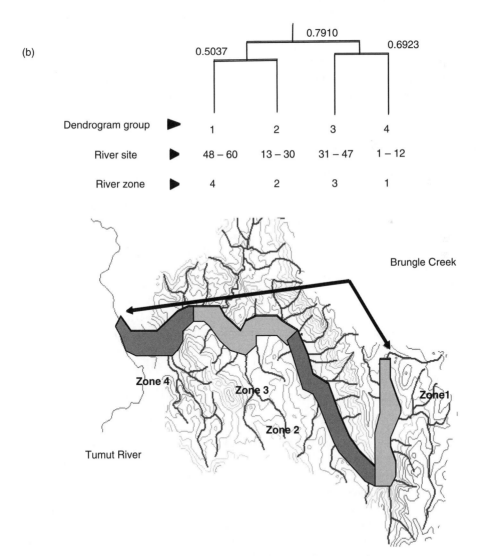

FIGURE 5.10 River zonation in Brungle Creek, Australia: (a) zones determined by a bottom-up process using cross-sectional data for 60 sites and (b) zones determined by a top-down approach as outlined in the text. Both approaches produced the same zonation pattern in this catchment.

Creek. The top-down approach to identifying process zones was undertaken at exactly the same 60 sites. The resultant dendrogram revealed four groups of sites (Fig. 5.10b) with similar catchment, valley, and channel conditions. This grouping of sites using the larger-scale variables produced the same configuration of FPZs within Brungle Creek as the moving window analysis of the smaller-scale, bankfull cross-sectional variables.

A Mantel test of the two distance matrices generated for the larger-scale variables and the smaller-scale bankfull cross-sectional variables revealed no statistical differences. A Mantels test is widely used for assessing the relationships between two distance matrices or, more generally, two resemblance or proximity matrices – in this case, two matrices determining FPZs. The test involves measuring the association between the elements in two matrices by a suitable statistic and then assessing the significance of this statistic by comparison with the distribution found by randomly reallocating the order of the elements in one of the matrices.

These analyses raise two important points. First, FPZs can be identified from various data and methods. Two different approaches and data sets provided the same distribution of FPZs in Brungle Creek: a top-down approach, which uses larger-scale catchment, valley, and channel planform variables, and a bottom-up approach, which uses smaller–scale, bankfull cross-sectional variables. Second, the scale of FPZs integrates both the constraint and processes operating at higher-level influences – catchment and valley scales – as well the influence of smaller-scale process responsible for shaping the character of the bankfull river channel.

SOME COMMON FUNCTIONAL PROCESS ZONES

A Brief Review of Functional Process Zones

Given the scale relevance of FPZs for river networks, the following section provides basic descriptions of those that have been encountered. As a reminder, FPZs are large sections – hydrogeomorphic patches – of the river network, which have similar catchment constraints (factors such as catchment geology, precipitation, valley confinement, and geological history), driving processes (discharge and sediment transport regimes), and larger-scale channel morphologies (usually depicted from their channel planform morphology). Channel planform is usually differentiated on the basis of three interrelated criteria, namely the number of channels, their sinuosity, and their lateral stability. The number of channels is commonly differentiated into large sections of rivers in which channels are: (i) absent or discontinuous and (ii) single or multiple. Rivers that are limited by capacity or competence normally have multiple channels and tend to have braided or wandering style planforms. In those parts of the catchment with low slopes and where the drainage breaks down, one encounters ana-branching, anastomosing, or distributary networks. Similarly in those regions of low energy where river systems are unable to incise into their valley floor, channels may be discontinuous or indeed absent. Sinuosity, defined as the ratio of channel length to valley length, typically varies between 1.0 (straight) and 3.0 (tortuous). Meandering rivers, which are normally defined as those with a sinuosity greater than 1.3, have many variants, and this is dictated by the slope of the river, the texture of riverbed sediments, the type of meander growth, and their behavior or growth. As a result, meandering channels tend to be more sinuous in regions of lower slopes or when flowing through more cohesive sediment. Alluvial channels, those that flow through their own deposits, are prone to lateral adjustments. The lateral stability of river channels is defined as the capacity of a river to adjust its position in the valley floor trough, and components of lateral stability include meander growth and shift along with the degree of braiding and avulsive behavior. Collectively, these various measures provide a means to characterize various types of FPZs.

The list provided below is by no means exclusive, and we expect river scientists to add to this list (see the comments section at the end of this book). These zones are grouped here by the degree of valley confinement.

Confined Valley Functional Process Zones

Confined Coarse-Textured Zone

In bedrock confined valleys, large amounts of coarse sediment can accumulate because of major erosional events, especially in those areas where valley confinement increases. In these areas, floodplain pockets develop, which are typically composed of coarsely textured sediments like cobbles and boulders. Hence, the direct influence of the adjacent valley slopes is decreased. Irregular in-channel accumulations of sediment form an array of sediment bars of different sizes and configurations. The variable bedrock influence has a marked effect on the diversity of physical structures associated with channels in this zone (Fig. 5.11).

FIGURE 5.11   Confined coarse-textured zone of the Chishui River, China. Photograph by Martin Thoms.

Gorge Zone

Gorges are deep, narrow valley sections of a river dominated by steep bed slopes (Fig. 5.12). They have been referred as canyon systems (Schumm, 1988) in which the valley sides are nearly vertical and the river channel occupies the entire valley floor. As a result, sediments are added directly to the channel from the adjacent valley slopes. Typically these high-energy sections are formed following headward retreat of large knick points, usually through areas of weakened rock. Gorges generally lack large accumulations of sediment and floodplains because of the associated high energies. Most gorges tend to have straight channels, although some may have a meandering pattern superimposed.

Headwater Zone

This FPZ is represented by the typical streams that many associate with the steeper headwater regions of a river network. Generally they are characterized by steep bed slopes, low

FIGURE 5.12   Gorge zone of the Zambezi River, Zimbabwe, Africa. Photograph by Martin Thoms.

FIGURE 5.13   Headwater zone of the Cotter River, Murrumbidgee catchment, Australia. Photograph by Martin Thoms.

width-to-depth ratios, and a stepped bed profile. Because of their high sediment transport capacity, most headwater zones function as sediment supply areas, which rapidly convey sediment downstream (Fig. 5.13). In many instances, the physical structure of headwater zones is heavily dependent on the structure of the surrounding catchment, and minor increases in the local configuration of the valley floor may be associated with areas of significant sediment accumulations.

Partially Confined Functional Process Zones

Boulder Zone

This FPZ usually forms in areas downstream of confined zones where local valley widths increase, thus allowing floodplain formation. They are high-energy systems because of the associated steep slopes (Fig. 5.14). Flows can occur in either a single- or a multichannel system, and channels are normally straight in planform.

Pocket Floodplain Zone

In valleys that meander through easily erodible rocks, discontinuous floodplains can form in those areas where valley floor widths increase, and there are major constrictions behind which sediments accumulate (Fig. 5.15). In those sections where the bed slopes are relatively high and the bed sediment is composed of coarse material, the riverbed sediment develops a characteristic armored layer, which is a distinguishing feature of these zones. Armoring, as noted in Chapter 4, refers to the development of a surface layer that is coarser than the sediment beneath it, and this coarse surface layer protects the finer materials beneath.

Mobile Zone

A mobile FPZ refers to that section of the river network that has a relatively active river channel in terms of the mobility of the riverbed sediment, and there are large quantities of sediments in temporary storage within the main active channel (Fig. 5.16). This sediment can be easily moved into transport during periods of higher than normal flows. These zones typically

FIGURE 5.14   Boulder zone of the Upper Murrumbidgee River, Australia. Photograph courtesy of the CRC for Freshwater Ecology.

have well-developed inset floodplain features and various in-channel bar systems, and the flood-plains are active in terms of the residence times of the stored sediment. Mobile zones occur downstream of confined sections of the river network where valley floor widths increase resulting in freely meandering but irregular channel planforms.

Pool Zone

This FPZ is normally located in headwater regions of the river network where valley widths increase and slopes decrease, thereby allowing the accumulation of alluvium. Pool zones can have small flanking floodplains with relatively narrow valley floors that limit floodplain development. Channels in the pool zone are characterized by small constrictions and the formation of significant backwaters or pools upstream of the constrictions (Fig. 5.17).

Unconfined Functional Process Zones

Anabranching Zone

This FPZ is associated with sections of the river network where sediment loads and flow energy promote the operation of multiple river channels separated by areas of floodplain

FIGURE 5.15   Pocket floodplain zone of the Maranoa River. Photograph by Martin Thoms.

FIGURE 5.16   Mobile zone of the Castlereagh River, Australia. Photograph by Martin Thoms.

(Fig. 5.18). These low-energy sections of the river network are characterized by stable, multi-channeled, suspended load systems (Nanson and Knighton, 1996). Anabranching zones with highly cohesive boundary sediments have been termed anastomosing channels, and here channels have very low width-to-depth ratios and feature high lateral stability. Normally associated with low stream energies, high-energy forms of anabranching zones can also occur.

FIGURE 5.17   Pool zone of the Upper River Murray, Australia. Photograph by Martin Thoms.

Bedrock Zone

In some unconfined settings, channels may flow over relatively thin accumulations of alluvium covering a bedrock base. Moon *et al.* (1997) described various single- and multi-channeled, bedrock macrochannels where the prominence of bedrock dictates highly variable channel morphologies (Fig. 5.19).

Braided Zone

This section of a riverine ecosystem is dominated by coarse bed load sediment transport in which bars, of many different forms, are constructed and thalweg shifts occur within multi-channeled settings. Additionally, the flow regime tends to be highly variable, and as a consequence, the number and location of channels and braid bars may change quickly (Fig. 5.20). The formation of braided channels is promoted by erodible banks, an abundant supply of sediment, and the rapid and frequent variations in stream discharge.

Discontinuous Zone

Channels typical of this FPZ are dynamic in time and space because of cycles of erosion and deposition, despite being associated with very low slopes and stream energies. During depositional phases, channels are virtually nonexistent, and flows are dissipated across a relatively intact alluvial surface; however, when erosion occurs, flows are confined to a single channel. Channels that flow through swampy meadows or bogs are typical of this form of river zone.

Distributary Zone

This FPZ is distinguished by a series of channels that separate or bifurcate repeatedly from each other and that may or may not join some distance downstream. The secondary channels persist relatively independently of the main channel from which they originate. Individual channels generally display a highly meandering form with low width-to-depth ratios, and the boundary sediments are normally composed of very fine sand, silt, and clay-sized particles (Fig. 5.21).

FIGURE 5.18 Anabranching zone of the Macquarie River, Australia. Photograph by Martin Thoms.

FIGURE 5.19   Bedrock zone of the Olifants River, South Africa. Photograph by Martin Thoms. (See color plate 6).

Floodout Zone

In regions of very low slope and high evaporation, channels may terminate in a floodout or terminal fan (Fig. 5.22). In these settings, the river system is unable to maintain a continuous channel or transport its sediment load; and beyond the channel sections, water and sediment are dispersed over an alluvial plain. Thoms and Rayburg (in press) provide a detailed account of the character of channels within a terminal fan.

Low-Sinuosity Zone

Straight rivers are generally low-energy, suspended load systems with channel sinuosities less than 1.3. The finer texture of the boundary sediments promotes low-capacity channels that are relatively uniform in cross-sectional shape. The associated floodplains are vertically accreted surfaces that have back swamps and low-level levee systems (Fig. 5.23). The floodplains may also contain a diversity of landforms, including scrolls, swales, in-filled channels, and abandoned surfaces of various forms.

Meandering Zone

This FPZ is a single-channeled system with a sinuosity less than 1.3 that generally form on moderate-to-low slopes and are associated with well-developed floodplains. Their channels typically migrate laterally across the valley floor or translate downstream (Fig. 5.24). Some river scientists distinguish between active and passive meandering systems (cf. Richards, 1982), with the former having ongoing bed and bank adjustments with well-developed riffle–pool sequences that are characteristically spaced at 5–7 times the width of the bankfull channel. Passive meandering refers to a situation where the meandering process is not fully adjusted, and as a consequence, irregularities within the planform are noticeable.

Wandering Zone

Channels in this zone are an intermediate form between meandering and braided channels and, therefore, have characteristics of each. Typically they have been observed between

FIGURE 5.20   Braided zone of the Godley River, New Zealand. Photograph by Martin Thoms.

FIGURE 5.21   Distributary zone of the Narran River, Australia. Photograph by Martin Thoms.

FIGURE 5.22   The floodout zone of the Lower Balonne River, Australia.

FIGURE 5.23   The low-sinuosity zone of the Lower River Tay, Perthshire, Scotland. Photograph by Martin Thoms.

meandering and braided zones, and Brierley and Hickin (1991) suggested that they represent a transition in slope and bed material texture. Wandering rivers tend to have fewer channels and active bars than those in braided zones. Normally a dominant channel can be identified, and this can be similar in character to that observed in meandering systems with a laterally active channel (Fig. 5.25).

FIGURE 5.24 The meandering zone of the Athabasca River, Canada. Photograph Courtesy of the Society of Economic Paleontologists and Mineralogists.

FIGURE 5.25 The wandering zone of the Lower River Murray, Australia. Photograph by Martin Thoms.

SUMMARY

Hydrogeomorphic patches and their resultant processes play important roles in riverine landscapes. They provide physical habitat and act as ecological disturbances, among other things. Therefore, gathering information on the physical characteristics is an essential first step in getting to know your riverine landscape, and this should be done at different scales. Systematic approaches to the characterization of larger-scale river networks have not received the same attention as those developed for smaller-scale river reaches or even those focused at the site scale. This is the result of a legacy of reductionist approaches dominated by the view that we can improve our knowledge of riverine landscapes and the systems that operate within them only with more and "better" measurements (Phillips, 1999). River networks are complex systems that need to be interpreted within their local and historical context. Attention to identifying and characterizing FPZs within the context of river networks will improve our knowledge of riverine landscapes. At this scale, river networks are a mosaic of patches that do not simply reflect a continuum of river zones.

In this chapter, we have delved into the messy business of methods – the how to's of determining the character of river networks. Characterization simply provides one of the various tools that can be applied to particular problems or contexts – it is not a substitute for focused observation, study, and clear thinking about the multitude of influencing processes. We caution against a focus on seeking the single method that can be applied to any situation. Even if this miracle method was thought to be found, it would represent the equivalent of iron pyrites – the fools' gold of the mineral world at best. What approach or tool to employ in characterizing river networks will depend on a number of things, including what data are available. The options presented in this chapter provide various approaches that can be applied to particular situations – but, they are not a panacea. As we have demonstrated here, FPZs can be identified equally as well via data collected at larger or smaller scales. Characterizations that highlight specific aspects of the linkages between channel networks and resultant processes are likely to be most useful, but careless application of any characterization may prove misleading – no characterization can substitute for an alert, intelligent, well-trained observer. Nonetheless, it is difficult to fully understand the properties of an FPZ or a collective of zones without reference to how they are defined by their catchment context, confinement, position in the network, and disturbance history as well their flow and sediment regimes and channel morphology. Consideration of these factors within a spatial hierarchy can further guide interpretations of field observations and evaluation of conditions within FPZs. Using the principles outlined in this chapter, FPZs can be identified and characterized in an objective and systematic manner, thus adding to our knowledge of riverine landscapes at multiple scales.

Ecological Implications of the Riverine Ecosystem Synthesis: Some Proposed Biocomplexity Tenets (Hypotheses)

Introduction
Distribution of species
Community regulation
Ecosystem and riverine landscape processes

INTRODUCTION

As described in Chapter 1, a major component of the RES is an expandable set of 17 biocomplexity tenets or hypotheses. We developed these model tenets both to illustrate some implications of the RES to broad areas of riverine ecology and to offer possible future research avenues in riverine ecology from headwaters to great rivers and throughout the riverine landscape.

The 17 biocomplexity tenets presented in this chapter describe the predicted functioning of epigean portions of riverine ecosystems on ecological timescales from species to landscape levels. The list is not meant to be exhaustive, and we hope our colleagues will add and test additional tenets derivable from the RES. *We make no claim to originality for all these tenets.* Some of these ideas are well-supported in the scientific literature, whereas others – to be perfectly honest – border on being educated guesses and could generate scientific debate. We made an effort to frame most tenets as testable hypotheses, at least in part, but this was easier for some ecological levels and topics than others. However, in most cases, more specific (spatiotemporally limited), testable hypotheses can be derived from these broader model tenets. We integrated our ideas about FPZs and ecogeomorphology into these tenets where they were applicable.

Our model tenets are divided into three somewhat arbitrary and overlapping categories. The first set of tenets concerns factors influencing species distributions or, in effect, composition of the species pool within a single riverine ecosystem. The next category on community regulation relates to factors controlling species diversity, abundance, and trophic position within the assemblage of species potentially present in the environment. Both density-independent and dependent factors are included. The final cluster of tenets covers processes at the ecosystem and riverine landscape levels.

All our model tenets are based on the assumption that riverine ecosystems around the world are pristine or at least have not been modified sufficiently to alter their basic community properties and ecosystem functioning. Clearly, this is rarely the case, at least in economically developed and developing countries. Therefore, we wrote Chapter 7 to explore how conditions in altered riverine ecosystems might influence predictions of these biocomplexity tenets.

All authors of scientific models naturally prefer to be right in their predictions, and we are no exception; however, our primary purpose here is to develop a research agenda to test some or all these ideas and to determine not just which hypotheses are false but *why* they may be true for some riverine ecosystems but not for others. In this manner, the accumulation of data from geographically and ecologically diverse riverine ecosystems will takes us that much closer to revealing truths about the general functioning of riverine ecosystems.

DISTRIBUTION OF SPECIES

The composition of the species pool in any given FPZ and riverine ecosystem is determined by chance events and many abiotic and biotic factors operating over timescales ranging from recent to geomorphic (e.g., historical position of glaciers in the north temperate zone and access to specific river channels) to evolutionary. Our tenets are more temporally constrained, however, with biocomplexity tenets 1–4 focused on ecological responses to current conditions, mostly along a river's longitudinal dimension. These emphasize the importance of habitat for individual species, whereas habitat effects at the community level are discussed in tenets 5–9.

Most of the model tenets described below relate to the longitudinal dimension of a riverine ecosystem because our physical model focuses more on this dimension, but the majority of hypotheses have applications to the lateral dimension. However, only model tenets 9, 12, 15, and 16 explicitly include the lateral dimension in name or process. We have intentionally omitted tenets on the vertical dimension because none of us currently has extensive research experience in this area. Instead, the reader may wish to consult various specialty articles on hyporheic ecology (see Chapter 2), including Boulton *et al.* (2002) and Poole *et al.* (2006). However, we suspect that densities and taxonomic richness within a specific hyporheic patch will be directly related to the degree of its hydrologic connectivity to surface waters. Richness and density should rise with an increase in oxygen concentrations and both the abundance and the lability of organic carbon (Fisher *et al.*, 2006). Particle size of the hyporheic zone should also significantly affect species distributions (Boulton *et al.*, 2002).

Model Tenet 1: Hydrogeomorphic Patches

> Species distributions in riverine ecosystems are associated primarily with the distribution of small to large spatial patches formed principally by hydrogeomorphic forces and modified by climate and vegetation.

This model tenet emphasizes the patchiness of species distributions (see also tenet 7) in response to the patchiness of the habitat at multiple scales. Habitat patches are formed in response to geomorphic structure, water presence and movement (hydrology and hydraulics), watershed vegetation (especially the riparian zone), and the interacting effects of climate. This patchiness varies over time and includes spatial scales ranging from millimeters to stream units (e.g., riffles) to at least the level of FPZs. Species perceive patches based on both the hydrogeomorphic characteristics of the habitat and the taxon's size and motility.

Both discharge patterns and geomorphic structures of the drainage basin and channel vary along a longitudinal dimension of the riverine ecosystem and can differ substantially among FPZs. The hydrologic nature of a FPZ relates to patterns of current velocity and discharge, with the latter including flow and flood pulses (<1 year), histories (1–100 years), and regimes (>100 years). Because the temporal signature of flow disturbance varies as you move downstream, the probability of occurrence of a particular FPZ in a given region of the riverine ecosystem may be enhanced. As you move downstream, however, the same type of FPZ may appear repeatedly in a complex pattern, which becomes less predictable above the ecoregional level. The unique habitat template characteristic of each hydrogeomorphic patch limits the potential species pool and alters biocomplexity within the associated FPZ (see also tenet 5). Absolute and relative abundance of species within FPZs is related primarily to the physicochemical nature of the FPZ and constituent species.

Model Tenet 2: Importance of Functional Process Zone Over Clinal Position

Community diversity and the distributions of species and ecotypes from headwaters to a river's mouth primarily reflect the nature of the functional process zone rather than a clinal position along the longitudinal dimension of the riverine ecosystem.

Tenet 2 is closely linked to the first hypothesis and represents a strong contrast to the clinal perspective, which has dominated ecological thought and research on streams since 1980 (see Chapter 2). This hypothesis suggests that the nature of the biota along a longitudinal dimension of a riverine ecosystem responds directly and primarily to the type of FPZ. Consequently, species distributions should be patchy at the FPZ spatial level, and biotic communities in comparable types of FPZs should be more similar to each other than either is to adjacent assemblages in different types of FPZs. A similar viewpoint was expressed by Poole (2002). This contrasts sharply with a clinal perspective of species distributions where a relatively smooth transition of taxa (a continuum) is predicted to occur along a longitudinal dimension and where community dissimilarity should increase directly and predominately with separation distance.

Various authors have concluded that patterns in species diversity in a riverine ecosystem bear little relationship to stream order (e.g., Statzner, 1981; Minshall *et al.*, 1982; Statzner and Higler, 1985; Townsend, 1989). Although we see some evidence of gross patterns (i.e., at a coarse spatial scale) of diversity and abundance for broad taxonomic groups along a longitudinal dimension (e.g., for fish, mollusks, crustaceans, and some insects), these patterns are highly variable and probably more responsive to the changes in types of FPZs that characterize the downstream progression. However, it is possible that the landscape arrangement of FPZs along a longitudinal dimension significantly influences biocomplexity (see tenet 17).

One of the primary tenets of the major clinal theory, the RCC (Vannote *et al.*, 1980; Minshall *et al.*, 1985), was that an important ecotype – functional feeding groups – changes in a continuous, predictable gradient downstream in response to changes in the allochthonous and autochthonous food sources that were considered most important for each region of the river. This clinal pattern has been challenged by some authors, such as Townsend (1989) who noted that the idealized downstream pattern of the RCC in primary trophic resources "... is remarkable primarily because it is not usually realized and cannot provide a worldwide generalization." However, the fundamental concept that primary types of functional feeding groups in a given reach should reflect the abundance, quality, and availability of organic matter is not controversial. Aside from arguments on the actual relative availability of organic matter in channel and slackwater habitats from headwaters to the mouth of a riverine ecosystem (e.g., Junk *et al.*, 1989; Sedell *et al.*, 1989; Thorp and Delong, 1994, 2002), the question some aquatic ecologists have raised is why the pattern of feeding groups does not reflect a continuous gradient. Some past proponents of a clinal perspective contended that tributaries in natural rivers and dams in altered rivers (Ward and Stanford, 1983b) serve as reset mechanisms that temporarily (spatially) interrupt this feeding gradient. We agree that the preponderance of different functional feeding groups at any site within an undisturbed river reflects the availability of high- to low-quality food, as determined by a large number of factors in the riverine landscape. However, we contend that this availability is determined primarily by the type of FPZ at that site rather than by the position along an alleged continuous, longitudinal gradient of physical characteristics of that river.

Model Tenet 3: Ecological Nodes

> Species diversity is maximum at or near ecological nodes representing hydrogeomorphic transitions between and within functional process zones and in areas of marked habitat convergence (e.g., tributaries) and divergence.

Most ecology textbooks report positive links between species richness and habitat complexity for many terrestrial and aquatic habitats. Consequently, we should expect to find the greatest diversity – all other things being equal – in areas of the riverine ecosystem where the complexity of the physical habitat, and possibly food, is the greatest. From a clinal perspective, the maximum complexity in the *main channel* has been thought to occur in medium-sized rivers at stream orders near four where the bed may still be coarse (in systems with high-relief headwaters), the riparian canopy does not cover the entire stream (promoting phytoplankton growth), light penetrates to bottom across most of the stream (permitting periphyton growth), *P/R* ratios are maximum, and diel temperature pulses are greatest (see Fig. 2 in Vannote *et al.*, 1980). Although the idea that diversity peaks in mid-order rivers is widely accepted, we suspect that this could be a false assumption if the entire riverscape is included in a survey and certainly if the entire riverine landscape is considered. In that case, the maximum diversity should occur in FPZs with extensive lateral dimensions and would probably be located within higher-order sections of many riverine ecosystems, especially in large rivers with extensive floodscapes. In these areas, habitat complexity for the riverine landscape is complex spatially and temporally, as can be the availability of organic food sources.

The clinal perspective has supported the concept of a gradual transition in species richness roughly following a bell-shaped curve but skewed toward lower stream orders. This gradual transition in diversity was challenged soon after the publication of the RCC by Statzner and Higler (1985). They maintained that stream hydraulics linked to geomorphic changes in the basin and bed were the most important environmental factors governing zonation of stream benthos on a worldwide scale. Rather than a steady gradient of stream hydraulics postulated by continuum models, they identified discontinuities in habitat conditions where transition zones in flow and resulting substrate size were the critical determinants of changes in species assemblages. Minshall *et al.* (1985) modified the conceptual framework of the RCC slightly to include effects of "... climate and geology, tributaries, location-specific lithology and geology, and long-term changes imposed by man." In essence, these were viewed as *bumps in the path* (our terminology) of a gradual transition in riverine properties that allowed for local corrections in an otherwise broadly consistent pattern.

Scientists have increasingly emphasized the importance of transitions within channels to ecosystem parameters (e.g., tributary confluences, divergence and convergence areas in braided/anastomosing rivers, and vegetated islands; e.g., Thorp, 1992), between flowing water channels and slackwaters (e.g., Schiemer *et al.*, 2001a; Thorp and Casper, 2002) and between slackwaters and sub-bankfull areas, such as parafluvial ponds (e.g., Burgherr *et al.*, 2002; Karaus *et al.*, 2005). Minshall *et al.* (1985) proposed that the effects of the tributary on ecosystem parameters depended on where the tributaries entered the larger stream. Tributaries entering upstream of the point of maximum diversity should cause the community to reach its maximum point sooner, whereas tributaries entering downstream of the peak should increase the diversity by driving the system back (i.e., resetting it). Side channels, braided channels, and off-channel backwaters (seasonal) may have an effect similar to that of tributaries" (Minshall *et al.*, 1985). Hyporheic transition areas may also serve as a form of permanent or temporary ecological nodes (cf. Ward and Voelz, 1994). An implication of the RES, however, is that the importance of a tributary confluence depends primarily on whether it alters basic hydrogeomorphic characteristics of the receiving FPZ.

In summary, the two competing models described above focus on opposite factors as an explanation for longitudinal changes in species richness. The proponents of a clinal model would argue that a gradual cline in physical conditions is responsible for the downstream pattern, but this is interrupted periodically by hydrogeomorphic blips from tributaries, bed changes affecting hydraulics, and substantial changes in lateral spatial complexity. Alternatively, other scientists contend that it is the abrupt transitions themselves that are responsible for the apparent broad pattern of change in the diversity of the river. In the latter case, therefore, one could hypothesize that total richness within the riverine ecosystem (gamma diversity) would rise with the increase in spatial complexity resulting from greater numbers of hydrologic and hydraulic transition points. A corollary of this hypothesis could be that species richness should rise in riverine ecosystems with an increase in the diversity of FPZs and the number of downstream transitions from one type of FPZ to another.

A fundamental prediction of the RES is that species richness at a given river site is directly linked to the nature of the hydrogeomorphic patches at various spatiotemporal scales. Hence, we hypothesize that the larger the number of FPZs in a natural riverine ecosystem, the greater the overall diversity that will be present as a result of inherent increases in habitat complexity. This habitat complexity should peak in areas surrounding changes in stream hydraulics. Therefore, our model tenet 3 hypothesizes that diversity will rise at ecological nodes formed by transitions both between FPZs and within these hydrogeomorphic patches in areas of marked habitat convergence and divergence. Hence, gradual downstream gradients of species richness should not be apparent except at the larger valley- or basin-wide scales, which can partially obscure the effects of FPZs. Some large-scale riverine ecosystem studies are needed to test these hypotheses.

Model Tenet 4: Hydrologic Retention

Overall community complexity varies directly with the diversity of hydrologic habitats in a functional process zone and increases directly with hydrologic retention until other abiotic environmental conditions (e.g., oxygen, temperature, substrate type, and nutrient availability) become restrictive.

Rivers are obviously distinguishable from lentic environments by the persistence of directional (advective) water flow, but all riverine ecosystems – from those with simple, constricted channels to ones with complex, intertwined secondary channels and broad floodscapes – include ecologically crucial riverscape areas near shore, in side channels, or in backwaters where directional currents are low or nearly permanently absent and where turbulence is greatly reduced.

Some lotic species require moderate current velocities for either proper abiotic conditions (e.g., higher oxygen concentrations and cooler temperatures for salmonid fish), access to food (e.g., drifting CPOM or invertebrate prey for net-spinning caddisflies), or reproduction (e.g., dispersing propagules or keeping eggs in benthic nests free from silt and fungal infection). For example, in a literature survey of eight large rivers of North America and Europe, Galat and Zweimüller (2001) determined that 50–85% of the fish species required some riverine (defined there as non-floodplain) habitats to complete their life cycles and 17.5% of the fish were in the more restricted category of fluvial specialists. Interestingly enough, macrohabitat generalists composed 44–96% of the nonnative fish in those rivers, in part because of their ability to exploit floodscape habitats.

Although flowing water is an important environmental attribute for most stream organisms, water currents also directly displace species downstream, alter the size and stability of the stream bed, and increase hydraulic stress. Hydrologic retention areas, or slackwaters, are essential elements of riverine ecosystems ensuring the natural functioning of healthy rivers, as demonstrated repeatedly (e.g., see references in Sedell *et al.*, 1989; Thorp, 1992; Thorp and Delong, 2002; Hein *et al.*, 2005) and modeled by the inshore retention concept (Schiemer *et al.*, 2001a). Slackwaters function as refuges from hydraulic and hydrologic stresses and serve as nursery and/or high-productivity reservoirs for many fish, other vertebrates, plankton, benthic invertebrates, vascular macrophytes, and some periphyton (e.g., Humphries *et al.*, 2006; Casper and Thorp, 2007). Moreover, they retain food longer and cycle nutrients differently from faster-flowing channel habitats. Although access to these lateral areas is not necessarily essential for most fish species in some rivers, such as temperate rivers with aseasonal flood cycles, it usually greatly enhances overall fish recruitment in most rivers (Humphries *et al.*, 1999, 2002; Winemiller, 2004). Most research on the importance of hydrologic retention areas has involved relatively long-lived structures in the riverscape (e.g., forested islands and bays; Casper and Thorp, 2007) and the floodscape, but ephemeral islands can be important in rivers of some ecoregions, such as the U.S. Great Plains (Thorp and Mantovani, 2005), and in braided, gravel-bed rivers (Karaus *et al.*, 2005).

The degree and the ecological importance of hydrologic retention vary with the type of FPZ. Slackwaters in FPZs with constrained channels are limited to shorelines not directly impacted by the thalweg, banks of occasional islands or sand/gravel bars, relatively deep pools if present, and perhaps large, in-channel rocks or temporary wood snags. The abundance and the diversity of lateral, hydrologic retention areas increase concurrently with a rise in the geomorphic complexity of the riverscape. Hence, FPZs characterized by braided channels, anabranches, and extensive floodscapes have a greater diversity of hydrologic habitats than are found in FPZs with constrained or simple meandering channels.

Maximum community complexity should occur in the most hydrogeomorphically complex FPZs, all other things being equal (see tenet 8). This is consistent with the ecosystem size hypothesis (Cohen and Newman, 1988), which specifically related habitat heterogeneity and availability to species richness. Within the gradient of hydrologic retention, we predict that the overall community diversity will increase directly with hydrologic retention until other abiotic environmental conditions (e.g., oxygen, temperature, substrate type, and nutrient availability) begin to substantially restrict the number of individuals or species that can tolerate those conditions. For example, during the summer in the U.S. Great Plains, zooplankton numbers are greater in slackwaters of the Kansas River than in the main channel; however, as surface flow stops, temperatures begin to rise, and oxygen levels fall at night, the numbers of many Metazoa diminish (Thorp and Mantovani, 2005, and personal observations). We hypothesize that the maximum diversity will be skewed toward areas of low to medium-low current velocities rather than occurring at intermediate or higher-level velocities. The effects of elimination of slackwaters by levees are discussed in Chapter 7.

COMMUNITY REGULATION

Factors regulating community structure are likely to vary significantly over spatiotemporal scales and should be influenced by the type of FPZ, its position within a riverine ecosystem, and the nature of the surrounding watershed and ecoregion. The relative importance of density-dependent and density-independent factors could change seasonally (e.g., Ward and Uehlinger, 2003) and will certainly vary with taxonomic group. Nonetheless, we list below some general conclusions applicable across broad spatiotemporal scales.

Model Tenet 5: Hierarchical Habitat Template

The most important environmental feature responsible for ecological regulation of community composition is a hierarchical habitat template, as determined primarily by interactions between geomorphic habitat features and both short- and long-term flow characteristics.

Gaining some semblance of unanimity on the relative importance of factors regulating lotic communities is a major challenge, but the factor that would probably be ranked highest by most riverine ecologists is a hierarchically scaled, habitat template (cf. Frissell *et al.*, 1986; Hildrew and Giller, 1993; Townsend and Hildrew, 1994) (also called the habitat *templet*). This factor, or really nested series of factors, integrates responses of aquatic species over various spatiotemporal scales. It focuses on interactive effects in the riverine ecosystem of geomorphic features of the aquatic and terrestrial habitat and flow characteristics (magnitude, frequency, duration, timing, and rate of change of hydrologic conditions; see Poff *et al.*, 1997). The genetic traits of a species make an individual or species more or less suited to a given habitat, which along with biotic factors, dispersal challenges (distance, barriers, and abilities), and chance events influence the community composition of any section of a riverine ecosystem.

A problem with the theory of habitat templates is that the definition is vague in actual use within the aquatic literature and may be equivalent in its extreme to a species' *niche*, defined as an *n*-dimensional hypervolume (Hutchinson, 1958). Thus, part of this unanimity among stream ecologists cited above is a simple result of the habitat template concept being so broad that it encompasses most of the environment of the organism, that is, where it lives (the physical and chemical environment) and how that abiotic environment changes over time. We admit, therefore, that this makes our tenet not only somewhat trite but also difficult to refute as a hypothesis. The habitat template theory has been presented as being at odds with neutral theory. In fact, as discussed by Thompson and Townsend (2006), both concepts help to explain differences among species in distribution within riverine ecosystems to varying degree depending on the environment, trophic level, and species traits (especially those related to dispersal abilities).

For purposes of this tenet, we are restricting the habitat template to a slight degree by only referring to abiotic conditions in the environment, that is, the physical and chemical properties of the habitat along with hydrologic and hydraulic characteristics of the surrounding water. Even here, these broad characteristics are often interactive. For example, the substrate can alter both current velocities and hydrologic retention, while flow strongly influences substrate particle size for everything other than bedrock (and even that can be altered over a long-enough time period).

If this tenet is still founded on such an extensive range of niche attributes, how can it contribute both to understanding the riverine environments and to shaping future research? We hope this tenet will prove useful in achieving these goals in at least three areas, which are as follows: (1) ranking relative importance of control factors; (2) evaluating changes along a longitudinal dimension of the river; and, (3) predicting effects of some types of anthropogenic disturbances. First, the tenet emphasizes the relatively greater importance of the spatially and temporally varying physicochemical environment to species distributions and community composition compared to the effects of dispersal limitations and biotic factors, such as species interactions and food availability. [It does *not* suggest, however, that these other factors are unimportant to communities; rather, it places these factors as secondary to the more encompassing effects of the abiotic environment.] Second, the relative importance of various components of the habitat template in selecting for different species traits should vary longitudinally in a riverine ecosystem according to the hydrogeomorphic characteristics of the local FPZ. Finally, by appreciating implications of the first two components above, it should be clear that substantial anthropogenic disturbance to the habitat template (e.g., altering flow regimes, geomorphic complexity in multiple dimensions of the river, substrate characteristics, or water-quality characteristics) could significantly impact the health of the aquatic community and functioning of the entire ecosystem.

Of our two defined components of the habitat template, most aquatic habitat research during the twentieth century focused on the importance of substrate and chemical characteristics to aquatic species. For more information on this subject, consult almost any introductory text on stream ecology. More recently, scientists have investigated how the physicochemical habitat, including substrate, is altered by flowing water or the lack of flow.

Research on the significance of flow characteristics – the other major component of the habitat template in streams – has been especially prominent in the scientific literature within the last decade, but the relative importance of mean and variability of flow patterns has not been firmly established. Vannote *et al.* (1980) hypothesized in the RCC, "... that the structural and functional characteristics of stream communities are adapted to conform to most probable position or mean state of the physical system." In contrast, Palmer *et al.* (2000) and others have argued that environmental variability itself may be a controlling factor. Moreover, Arrington and Winemiller (2006) demonstrated that the degree of habitat specificity changed substantially with flow characteristics. It is possible, therefore, that any observed differences in biocomplexity among types of FPZs could be as much a response to hydrogeomorphic variability (e.g., fluctuations in flow rates and channel complexity from floods and droughts) within an FPZ or whole stream as a response to differences in the mean state of the environment. Given that variability of flow is inversely related to stream size within the same ecoregion, the importance of short-term hydrologic variability in community regulation might be greatest in headwater streams (but see tenet 6). In addition to the role of in-channel flow in controlling community structure and ecosystem function (e.g., nutrient spiraling), sub-bankfull flow pulses (Tockner *et al.*, 2000) and supra-bankfull flood pulses (Junk *et al.*, 1989; Winemiller, 2004) are critical to sustaining biodiversity, probably more so in downstream portions of the riverine ecosystem.

The relative and absolute effects of dynamic geomorphic features and hydrologic conditions on biocomplexity will vary with the nature of the FPZ. This reflects in part the observation that short- and long-term variability and predictability of environmental conditions differ among types of hydrogeomorphic patches both within and among rivers. The degree to which environmental variation becomes a disturbance or perturbation in the system is related to the range of variation and its predictability, both of which will vary among ecoregions, rivers within ecoregions, and FPZs within rivers. It is conceivable, therefore, that the relative importance of genetic traits linked to resilience and resistance to stress differs among FPZs according to the frequency and predictability of environmental stresses affecting habitat characteristics (cf. Townsend *et al.*, 1997; Woodward *et al.*, 2005).

Model Tenet 6: Deterministic vs Stochastic Factors

Both deterministic and stochastic factors contribute significantly to ecological regulation of communities, but their relative importance is scale- and habitat-dependent and is affected by the organism's life history characteristics vs timescale and degree of environmental fluctuations; however, stochastic factors are more important in general and throughout the riverine ecosystem.

Stochastic, nonequilibrium processes in riverine ecosystems not only seem to be associated primarily with flood and flow pulses (especially the former), droughts, and stream hydraulics (e.g., direct effects of water velocity and turbulence and indirect effects from substrate movement) but are also related to other environmental disturbances associated with weather events, watershed inputs, and possibly climate change over longer periods (e.g., increased intermittency; e.g., Dodds *et al.*, 2004). Weather impacts on water temperatures, oxygen content, and ice scouring are occasionally important, especially in smaller streams. Likewise, stochastic factors influencing the watershed, such as deforestation from fires, winds, and possibly herbivores, can be consequential. The spatial and temporal scales over which stochastic factors commonly exert significant impacts should vary with the nature of the disturbance and the organism or process examined, as illustrated by four diverse processes: rock tumbling, hydrological pulses, droughts, and watershed disturbances.

A stream spate, especially one in a headwater stream, will move bed material easily (e.g., sand to gravel), moderately well (pebbles and cobble), or with difficulty (boulders), creating the potential for stochastic disturbances over varying spatial and temporal scales.

Movement of sand may be frequent enough to become a normal feature of the environment, and thus nonstochastic. However, if the sand moves substantially, it can alter the presence of an ephemeral hydrologic retention zone, changing important aspects of the habitat such as temperature, oxygen, and organic content as a consequence of increased current velocity. Changing the distribution of silt, sand, and gravel in a river can also produce or eliminate regional dispersal barriers for benthic species, such as some pleurocerid snails in the Ohio River whose longitudinal distribution is restricted by silty substrates separating favorable habitats of gravel or larger rocks (Greenwood and Thorp, 2001). A spate large enough to overturn a cobble can alter interspecific space competition among herbivorous or net-spinning invertebrates, such as snails and caddisflies (Hart, 1983; Georgian and Thorp, 1992), or reduce/eliminate the general importance of both competition and predation (e.g., Lake, 2000). Fish may not perceive these changes in current velocity as disturbances, but spates large enough to transport large-particle bed material will likely influence most fish in the stream and serve as a stochastic stress.

Hydrological pulses vary in degree from sub-bankfull flows to supra-bankfull floods and in frequency from multiyear to subseasonal; their predictability in timing and length are major factors (along with hydraulic force) in determining whether the pulses are generally considered stresses or subsidies to the ecosystem. With a decrease in the predictability of hydrological pulses in time and degree (spatial extent and amount of hydraulic force), the stochastic nature of the pulses rises along with the extent of disruption to community structure and ecosystem processes. At the opposite extreme, the reduction in hydrologic pulses can itself be a stochastic disturbance if species have adapted to predictable floods. Evidence abounds that many but not all fish require a flood period for reproduction (Galat and Zweimüller, 2001; Winemiller, 2004) and growth (e.g., Balcombe *et al.*, 2005), and some species of riparian and floodplain plants, such as red gum trees (*Eucalyptus camaldulensis*) along the Murray River in southern Australia and cottonwood trees (*Populus deltoides*) along the Missouri River in Montana, USA, cannot flourish without periodic floods (e.g., Scott *et al.*, 1997; Robertson *et al.*, 2001; Overton *et al.*, 2006).

Droughts vary from being predictable to stochastic in their effects, and they are usually considered environmental stressors. Some headwater streams in all ecoregions and many rivers in dryland ecoregions have intermittent flow. Conditions range from minimal surface flow to formation of surface pools connected only by hyporheic flow, to elimination of all surface water. Such conditions clearly restrict the number of native and potential exotic taxa that can live in these environments, especially when the droughts are lengthy (Marshall *et al.*, 2006; Sheldon and Thoms, 2006b). However, variability in flow has been shown to be of conservative value in some systems (Sheldon *et al.*, 2002), and it is clear that periodic droughts favor some plant and animal taxa. Species thriving in rivers subject to droughts require dispersal, physiological, behavioral, and/or life history traits of resistance or resilience to droughts. Some examples are amphibious pulmonate snails in headwater streams of North America (Brown, 2001), burrowing lungfish in Africa, and many species of branchiopod crustaceans throughout the world with drought-resistant propagules (Dodson and Frey, 2001).

Aside from direct anthropogenic effects, watersheds are periodically disturbed by stochastic events that directly harm or destroy vegetation and indirectly alter inputs to the riverine ecosystem, usually over a large area. These watershed stressors could include, for example, forest fires, blowdowns, landslides, and rarely tectonic events, such as a volcanic eruption (e.g., Lamberti *et al.*, 1992). Fires, especially crown fires, can strip a valley of the trees and other vegetations that hold soil in place. This could easily result in negative ecological responses of stream communities to short- to long-term increases in runoff of water, sediments, inorganic nutrients, and organic matter into streams (Mihuc and Minshall, 2005; Shakesby and Doerr, 2006) as well as changing detrital constituents of the floodscape. Disruption of the vegetation and soil constituents of floodplains could substantially modify effects of the flood pulse cycle (personal communication, Prof. Wolfgang Junk). The watershed disturbance might alter, for example, the availability and chemical nature of nutrients stored on the floodplain and the aquatic "seed bank" of resistant eggs and embryos, which can be mobilized by floods. Not included in this list are stochastic factors linked to human actions altering either the nature of the watershed or direct inputs to streams (but see Chapter 7).

The frequency of environmental disturbance is often as important as the intensity, and these may be interactive characteristics. One of the most commonly cited theory related to frequency and intensity of disturbance is the intermediate disturbance hypothesis (IDH) of Connell (1978). Its use in riverine ecology has been reviewed by Ward and Stanford (1983a), Townsend *et al.* (1997), and Ward and Tockner (2001). The IDH has been applied to various abiotic (e.g., hydrologic and hydraulic) and biotic disturbances (mostly predator-prey) in small streams through large rivers. Unfortunately, it has occasionally been employed in a rather cursory fashion. The IDH was originally based on studies of marine benthos, especially competing species of barnacles and mussels along with their starfish predators (e.g., Paine, 1966). The basis of the concept was that species diversity was enhanced because the disturbance (predation originally but later abiotic disturbances such as log bashing; Dayton, 1971) interrupted processes of competitive exclusion between species with different colonizing and competitive abilities by producing bare space on intertidal rock faces. Connell (1978) broadened the concept's utility for multiple ecosystems, including tropical rain forests. Stream ecologists initially employed the model to look at deterministic processes, but a definitive demonstration of an intermediate biotic disturbance effect has rarely been attempted in freshwater (but see a lentic field experiment by Thorp and Cothran, 1984). Instead, the vast majority of riverine applications of this model have dealt with abiotic disturbances (e.g., Piscart *et al.*, 2005), especially hydrologic. In most cases, researches have seen a curvilinear response pattern at some spatiotemporal scale, and have then ascribed that to the IDH with little explanation or proof that the process was rooted in assumptions of this theory (but see Death, 2002). Wootton (1998) questioned the application of this theory to some complex, multitrophic ecosystems, such as lotic communities. For a thorough discussion of patterns of environmental perturbations (disturbances and responses), see Lake (2000).

Ecological research on effects of stochastic processes in streams continues (e.g., Uehlinger, 2006), but its intensity has clearly diminished even though many questions still remain. The following are a few research questions that need to be addressed. *Are stochastic factors more important at smaller spatiotemporal scales than larger ones?* One might surmise that processes operating at the scale of a cobble (e.g., rock turning) would be more subject to stochasticity rather than activities transpiring at the reach or higher scale. *How is the importance of spatiotemporal scale in this question influenced by species traits and the type of disturbance?* It seems probable that motile species are susceptible to fewer stochastic stressors than more sedentary species. *How does the importance of stochastic events vary with stream size and FPZ?* The first response to this question might be an assertion that communities in small streams and relatively geomorphically simple FPZs (e.g., sections of a river with a constricted channel) are more susceptible to stochastic events because: (i) discharge variability increases directly with mean discharge in the same ecoregion and (ii) lateral dampening of hydrologic pulses should increase with the geomorphic complexity of the FPZ. On further thought, one might hypothesize that native species in "stochastic environments" have adapted to or been selected for traits that ameliorate the effects of a stochastic stressor, perhaps to the point where it is no longer significant or stochastic. The same may be true for an analysis of stochasticity by ecoregion. That is, stream conditions probably fluctuate more or less predictably on average in drier climates, higher elevations, and temperate zones than in forested ecoregions, lowland areas, and high or low latitudes, but has evolution-designed species capable of handling (ignoring) much of this environmental variation? *What genetic characteristics of species best adapt them*

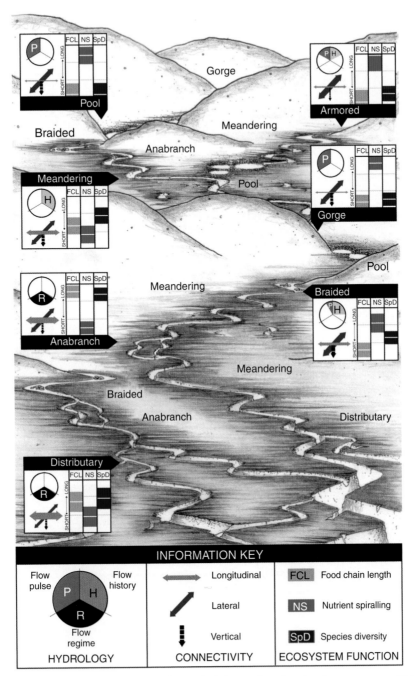

PLATE 1 A conceptual riverine landscape is shown depicting various functional process zones (FPZs) and their possible arrangement in the longitudinal dimension. Not all FPZs and their possible spatial arrangements are shown. Note that FPZs are repeatable and only partially predictable in location. Information contained in the boxes next to each FPZ depicts the hydrologic and ecological conditions predicted for that FPZ. Symbols are explained in the information key. Hydrologic scales are flow regime, flow history, and flow regime as defined by Thoms and Parsons (2002), with the scale of greatest importance indicated for a given FPZ. The ecological measures [food chain length (FCL), nutrient spiraling, and species diversity] are scaled from long to short, with this translated as low to high for species diversity. The light bar within each box is the expected median, with the shading estimating the range of conditions. Size of each arrow reflects the magnitude of vertical, lateral, and longitudinal connectivity (See Figure 1.1, p. 3).

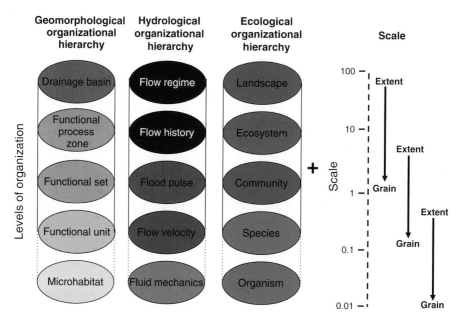

PLATE 2 Organizational hierarchies in river science. To use this framework, one must first define the relevant spatiotemporal dimension for the study or question. Scales for each hierarchy are then determined and allow the appropriate levels of organization to be linked. The scale at the right demonstrates that linking levels across the three hierarchies may be vertical depending on the nature of the question (See Figure 3.2, p. 24).

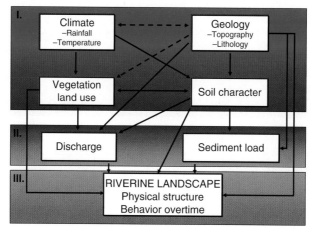

PLATE 3 Factors influencing the physical structure and behavior of riverine landscapes over time (modified from Morisawa and Laflure, 1979). Three levels of influence are recognized: (I) independent catchment factors; (II) independent channel variables; and (III) dependent channel variables as described by Schumm (1997a). (See Figure 4.1, p. 42).

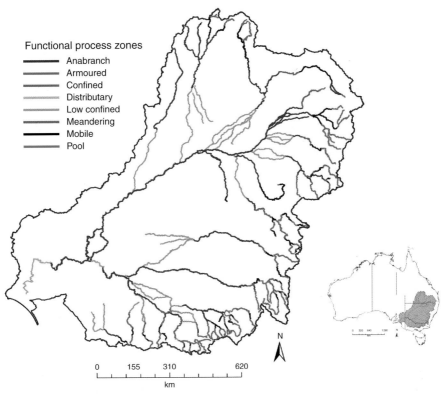

PLATE 4 The spatial distribution of FPZs within the Murray–Darling Basin, Australia, at the 1:500,000 scale (modified from Thoms *et al.*, 2007). Inset shows rivers of Murray–Darling Basin and location in Australia. (See Figure 4.4, p. 52).

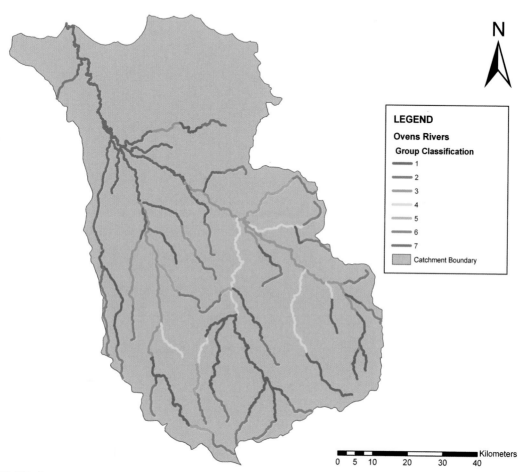

PLATE 5 River zones within Ovens catchment, Australia. These river zones were determined from a top-down approach with methods described in the text. (See Figure 5.5, p. 79).

PLATE 6 Bedrock zone of the Olifants River, South Africa. Photograph by Martin Thoms. (See Figure 5.19, p. 98).

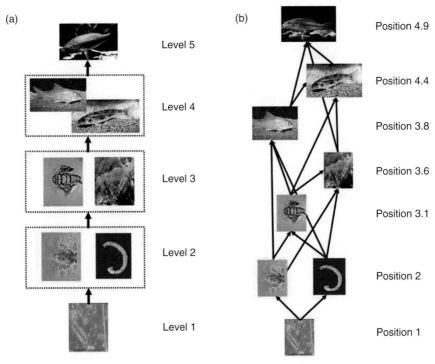

PLATE 7 Comparison of food chain length using (a) trophic-level approach and (b) trophic-position approach for a generalized food web. Fish images by Konrad Schmidt, autotrophs by John Wehr, and invertebrates from http://www.glerl.noaa.gov. (See Figure 6.1, p. 116).

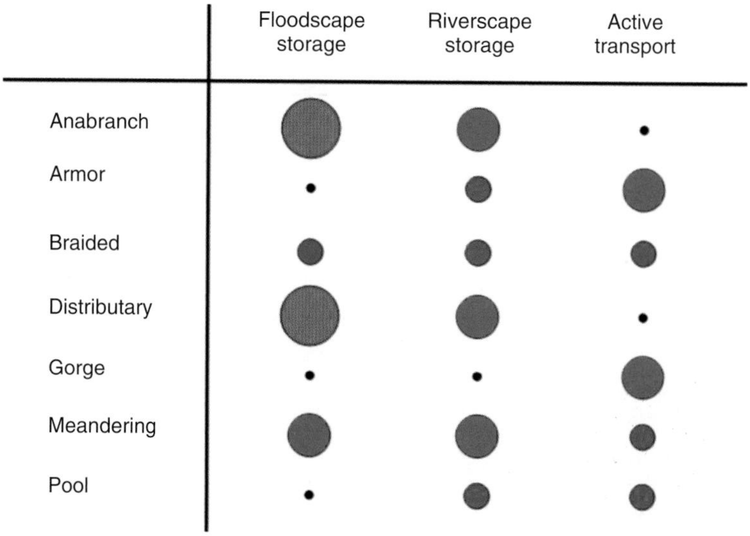

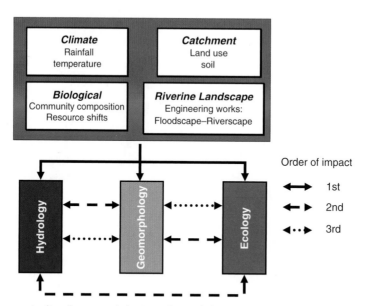

PLATE 8 Potential for nutrient storage in the floodscape and the riverscape and active downstream transport in representative FPZs. Size of circles represents the relative amount of nutrient retention and active transport. This can be used to define the length of nutrient spirals for FPZs in Fig. 1.1 (See Figure 6.2, p. 125).

PLATE 9 Anthropogenic disturbances and the immediacy of their influence on the three domains of the riverine landscape. Order of impact reflects if the effects are: (a) direct influence on a domain (first order); (b) indirect on the second domain through effects on an initial domain (second order); or (c) indirect through changes first imparted on two domains before influencing the third (third order). (See Figure 7.1, p. 134).

(a)

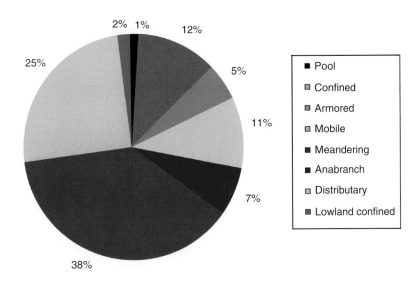

(b)

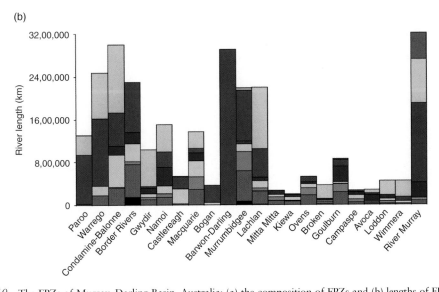

PLATE 10 The FPZs of Murray–Darling Basin, Australia: (a) the composition of FPZs and (b) lengths of FPZs within the various subcatchments of the Murray–Darling Basin. Modified from Thoms *et al.* (2007). (See Figure 8.1, p. 172).

for ecologically unpredictable disturbance and how do these traits vary with the type of stressor? These species traits could include the life span of the organism in relation to the frequency of disturbance, as discussed later. Finally, one might ask: *does the relative importance of stochastic and deterministic factors in regulating taxa and processes differ across a lateral dimension of a river (related to the question about FPZs)?* This last question is partially addressed below.

Up to this point in our discussion of tenet 6, we have focused on stochastic factors. This is justifiable in part because the prevailing view over the last several decades has been that rivers are nonequilibrium environments dominated by abiotic, advective forces (e.g., Pace *et al.*, 1992). However, deterministic factors may be important on some timescales and prevalent in some riverine habitats (Thorp *et al.*, 1998; Jack and Thorp, 2002; Thorp and Casper, 2003). In particular, deterministic factors are more likely to be significant in portions of the riverscape where water currents are minimal in absolute terms and relative to a species' swimming abilities.

Deterministic processes contributing to community regulation in riverine ecosystems are mostly related to interspecific competition, predator–prey interactions, and host–parasite interactions. The current evidence for interference (space) competition (e.g., Hart, 1983; Georgian and Thorp, 1992) seems stronger than for exploitative (primarily food) competition. Nonetheless, some studies have demonstrated that resource competition can influence population size in zooplankton (e.g., Pace *et al.*, 1998), snails (e.g., Brown, 2001), and other organisms. Deterministic processes tend to be more influential on shorter rather than longer timescales and occur where physical stress is diminished. Therefore, one would more likely find significant biotic interactions in: (i) hydrogeomorphically complex FPZs and (ii) lateral slackwater habitats, or perhaps in the main channel at times or in areas of minimal hydraulic stress. Although not as often reported as in lentic environments, trophic cascades of predator/herbivore and prey have been suggested for some lotic ecosystems (e.g., Pace *et al.*, 1999; Woodward and Hildrew, 2002; Thorp and Casper, 2003).

The significance of deterministic factors among and within FPZs is influenced by the degree of stress and the frequency of stochastic events. Advective and hydraulic forces in some rivers may be strong enough in the main channel even under normal flow conditions to prevent some species (e.g., many species of branchiopod zooplankton) from reaching densities where competition would be important. A severe stochastic event may also depress population densities so low that intra- and interspecific competition is unlikely throughout the riverscape, and thus the potential role of predators in regulating species diversity (Thorp, 1986) would be reduced.

Two other factors that could be very crucial in determining the relative role of deterministic and stochastic processes are the life span and the reproductive period of that organism. In his classic paper entitled *The Paradox of the Plankton*, Hutchinson (1961) argued that the likelihood of competitive exclusion was related to the relative times necessary for both exclusion and competitive-altering environmental change. In a similar vein, the relative importance of deterministic processes in riverine habitats may be related to the life span and/or the reproductive period of the organism in question vs the frequency of stochastic environmental disturbances in a particular habitat. If the environmental disturbances occur frequently during an organism's life span and are not too severe, the species may be adapted to these abiotic fluctuations, and deterministic processes could be relevant. Likewise, if the disturbance period is long in relation to a species' reproductive period or life span, deterministic processes could become prominent during the long intervals between major disturbance events, such as a major drought. Thus, stochastic factors are most likely to play a major role in the lotic community when either: (i) the disturbance is sufficiently strong no matter what its occurrence pattern or (ii) the frequency of environmental disturbance is comparable to either the life span or the reproductive period of the species affected.

One clear conclusion from these arguments is that community regulation by stochastic or deterministic factors is a strongly hierarchical process dependent on the spatial and temporal scales operating above and below the level of the observed pattern (see also Chapter 3).

Model Tenet 7: Quasi-Equilibrium

A quasi-equilibrium is maintained by a dynamic patch mosaic and is scale dependent.

A true equilibrial state is unlikely in riverine ecosystems because they are open, advective systems subject to major hydrologic variations, which introduce large measures of stochasticity among and within patches at various scales. Despite the likely predominance of stochastic, nonequilibrial processes at most spatiotemporal scales and habitats, we believe that lotic communities lacking significant anthropogenic disturbance are generally maintained in a quasi-equilibrial state. For such a state, one should expect to find approximately the same species over time, with only moderate shifts in relative abundance from year-to-year (baring major abiotic disturbance events or invasion by exotic species) and no major shifts over periods of decades. For systems regularly undergoing major fluctuations in flow, such as drought-prone rivers, the time period over which the quasi-equilibrial state functions should be longer than for continuously flowing riverine ecosystems. Because of this quasi-equilibrial state, stream naturalists can lead a field trip and have a reasonably good idea of which species they will find and in what general abundance, even though few would care to bet on the exact relative rank of species at any given time unless the faunal diversity is low.

As described in greater detail in Chapter 3, a quasi-equilibrial state should result from the incorporation of multiple, nonequilibrial or short-term equilibrial processes operating at lower hierarchical levels in a dynamic mosaic of patches characterized by multiple spatial and temporal dimensions (cf. Paine and Levin, 1981; O'Neill *et al.*, 1989). Put more simply, this means that patches are constantly changing but the processes occurring on one patch may be different in nature and temporal scale than what is occurring on adjacent patches. For example, two species of potential space competitors may be interacting on two adjacent rocks, one a small cobble and the other a large one. If a flow pulse turns the former over and not the latter, then rock-level competitive displacement may occur in the second instance but not in the first. If a member of the competitive inferior species is superior at colonizing another empty cobble space and later reproduces, then the balance of species is maintained. Thus, a summation of results from multiple patches behaving in slightly different ways combined with the effects of stochastic processes produces a semblance of equilibrium at the community level.

At a larger scale, a quasi-equilibrium might be maintained by metapopulations. A metapopulation is a spatially fragmented set of populations of the same species that are spatially isolated on shorter timescales but that exchange individuals and genes over a longer temporal period. This allows a population in one spatial unit to go locally extinct but have the species reappear later through colonization from a distant and often more favorable site. The study of metapopulations in open, advective systems is in its infancy, and it is too early to ascertain their importance for lotic populations. Defining a metapopulation for lotic communities, however, is somewhat risky. Does recolonization of intermittent streams reflect the action of a metapopulation? Likewise, does recruitment of fish and invertebrates into channel habitats from large lateral patches constitute metapopulational dispersal? One of the more likely examples of the effects of a metapopulation concerns recruitment of resistant potamoplankton propagules from one catchment to another via prevailing winds or waterfowl movement; however, the relative importance of this process has not been sufficiently demonstrated or even adequately addressed.

Model Tenet 8: Trophic Complexity

> Food chain length (maximum trophic position) increases directly with the hydrogeomorphic complexity of a functional process zone in response to multiple factors related to habitat heterogeneity, diversity and abundance of food resources, and dynamic stability.

Food chains and food webs have traditionally been based on the concept of trophic levels (e.g., producers, herbivores, first-order carnivores, second-order), which were treated as if they were constant and relatively inviolable even though the presence of omnivory is widely known to be common in many ecosystems, especially (we suspect) in riverine ecosystems. Trophic complexity could thus be measured by an increase in the number of trophic levels as well as by the number of interactions among species between levels. Species feeding at different trophic levels were considered to be in different functional feeding groups based on what they ate or sometimes how they obtained that food. For example, many snails feed on attached microalgae and are thus called scrapers (herbivores), while predatory water bugs may consume herbivorous mayflies or predatory stoneflies by sucking out their prey's body fluids and are therefore known as piercers (in these cases, a first- and second-order carnivores, respectively). This approach using functional feeding groups, although widespread and somewhat useful, has been roundly criticized for not properly accounting for the extent of omnivory or life cycle variations (e.g., Mihuc, 1997). Anyone who has tried to assign adult riverine fish to a single trophic level can appreciate the difficulty of this task for all but a few taxa, such as paddlefish (*Polyodon spathula*, a zooplanktivore) and freshwater gar (e.g., *Lepisosteus* spp., almost strictly piscivores as adults). Most riverine fish in the temperate zone could probably be categorized as omnivores if analyzed over long enough periods during the adult stage.

In contrast to trophic-level models, which provide a simple identifier of trophic level, *trophic-position* models recognize that consumers feed at multiple levels, thereby incorporating omnivory into the model. In contrast to fixed-integer values of trophic levels (e.g., $1 =$ producer, $2 =$ strict herbivore), trophic position is a continuous variable that changes in response to food quality, food availability, and degree of omnivory (Fig. 6.1). Trophic position can be measured quantitatively using carbon and nitrogen stable isotope ratios (depicted as $\delta^{13}C$ and $\delta^{15}N$, respectively; Post, 2002b; Fry, 2006; Anderson and Cabana, 2007). Trophic position has been used to examine contaminant bioaccumulation in food webs (Cabana and Rasmussen, 1994; Vander Zanden and Rasmussen, 1996), the implications of nutrient loading on trophic dynamics (Lake, 2003; Fry and Allen, 2003), and the impact of introduced species on food web structure (Vander Zanden *et al.*, 2003). Trophic position has also been used to identify differences in food chain length (FCL or maximum trophic position) as a function of productivity in Venezuelan rivers (Jepsen and Winemiller, 2002) and is being employed by two of us (Delong and Thorp, current EPA project, unpublished) to evaluate effects of major anthropogenic disturbances in rivers over periods from a few decades to more than 100 years.

Current models of FCL primarily emphasize the importance of either *energetics* (resource limits), *dynamic community stability*, or *ecosystem size*, though some have argued for multifactorial explanations involving these and other variables (e.g., Post, 2002a). In riverine ecosystems, one could evaluate factors affecting FCL with comparisons among and within rivers, with the latter particularly emphasizing FPZs. A prominent model on the role of resource limits is the productive-space hypothesis (Schoener, 1989; Vander Zanden *et al.*, 1999). Within riverine ecosystems, FPZs with more lateral complexity in the riverscape will include more areas of minimal currents (slackwaters) that should enhance the diversity and productivity of autotrophs, thus increasing the amount of productive space. Surprisingly enough, however, Jepsen and Winemiller (2002) found that top piscivores in nutrient-poor rivers had higher trophic positions than those in more productive rivers (but see discussion in Chapter 7, tenet 8). Moreover, Thompson

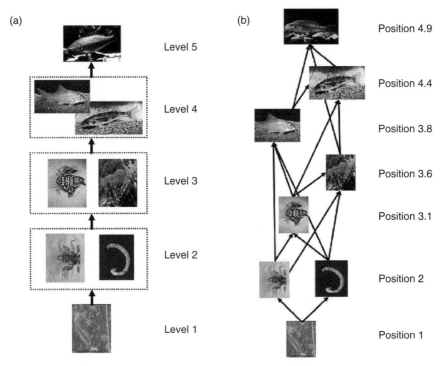

FIGURE 6.1 Comparison of food chain length using (a) trophic-level approach and (b) trophic-position approach for a generalized food web. Fish images by Konrad Schmidt, autotrophs by John Wehr, and invertebrates from http://www.glerl.noaa.gov. (See color plate 7).

and Townsend (2005b) reported a positive correlation between algal productivity and FCL in streams passing through grasslands but found no relationship in forested sites (where FCL was instead related to ecosystem size). An analysis of generically similar ecological studies led Post (2002a) to conclude that resource availability limits FCL only in systems with very low rates of productivity (such as $<10\,g\,C/m^2$ per year). Another factor potentially affecting FCL in riverine ecosystems is dynamic stability, which is influenced by both hydrologic patterns (variability of discharge and size of flood pulses) and geomorphic complexity in space and time. Therefore, it should differ among rivers and within riverine ecosystems by type of FPZ; hydrologic connectivity within the riverine landscape should also alter dynamic stability (e.g., Roach, Thorp, and Delong, unpublished data). Ecosystem size, which is probably now considered the primary regulator of FCL (e.g., Post *et al.*, 2000), naturally varies among rivers. But it also could be evaluated, in a sense, within riverine ecosystems by examining the effects of complexity, lateral extent, and longitudinal size of FPZs. Ecosystem size is an aggregate variable of a number of biotic and abiotic factors, including habitat heterogeneity, which varies directly with FPZ complexity. A greater diversity and amount of habitats and food resources could promote a general rise in dietary specificity, a relative drop in omnivory, and an increase in FCL (cf. Post *et al.*, 2000; Woodward and Hildrew, 2002; Woodward *et al.*, 2005). For example, with increases in habitat heterogeneity, species richness, and prey abundance, more opportunities would be available for adult "piscivorous" fish to feed exclusively on other fish and to do so earlier in their life cycles (once they are not gape or behaviorally limited to other prey).

Therefore, FCL within riverine ecosystems seems directly related to the hydrogeomorphic complexity of the local FPZ as a result of multiple factors advocated by general FCL models in ecology.

Model Tenet 9: Succession

The relative importance to multicellular organisms of different types of succession (simple species replacement, true facilitative succession, and true non-facilitative succession) varies between the riverscape and floodscape, among functional process zones, and across gradients of hydrologic variability. Classical (facilitative) succession is primarily limited to terrestrial elements of the riverscape (on ephemeral islands) and the floodscape (riparian and floodplain habitats), and it occurs principally in response to hydrogeomorphic processes. Simple species replacement increases its importance over true succession (facilitative and non-facilitative) as hydrological variability decreases in the wetted portion of the riverscape.

Research on succession in riverine ecosystems has until recently been considered mostly an esoteric subject, but now the role of succession in riverine landscape rehabilitation is of increasing interest (e.g., Webb and Erskine, 2003; Van Geest *et al.*, 2005; Baptist *et al.*, 2006). Unfortunately, scientists have rarely addressed how the nature of succession varies with habitat type at various spatial scales within a riverine ecosystem, and basic confusion abounds because of imprecise use of the term *succession*. Classical succession, or *facilitation* (described in Connell and Slatyer, 1977), is probably limited mostly to terrestrial elements of the riverscape (bars and islands) and the riverine landscape and results from hydrogeomorphic processes operating on medium timescales of one to hundreds of years. Some examples of riverine landscape succession are given in Robertson and Augspurger (1999) and Friedman and Lee (2002) and are also discussed in general in Piegay *et al.* (2000) and Amoros and Bornette (2002). Primary, facilitative succession principally entails predictable vegetative seres both on initially bare alluvial islands within the active river corridor of the riverscape and on bare alluvium deposited in the floodplain following episodic floods (cf. Ward *et al.*, 2002). Terrestrial succession in these areas significantly impacts biocomplexity in the linked aquatic ecosystem. Newly formed floodscape lakes within the riverine landscape will also undergo succession, which can ultimately lead to terrestrialization unless reset by a new flood; Ward *et al.* (2002) refers to this as hydrarch succession. Long-term hydrogeomorphic processes lasting thousands of years or more will also gradually shift characteristics of the channel and lateral habitats causing development of new FPZs and successional communities.

Although facilitative succession among multicellular assemblages probably rarely, if ever, occurs in wetted portions of the riverscape, many authors have concluded that other forms of succession are present. In most cases, this putative succession may just be annually repeated, seasonal cycles of *species replacements* or perhaps the simple cumulative effects of dispersal into a previously disturbed environment to fill empty niches (as predicted by island biogeography theories even without facilitative succession). In other instances, however, an interaction of species over time and/or a response to changing environmental conditions following a hydraulic disturbance will cause a predictable shift in ecosystem processes (e.g., Fisher *et al.*, 1982) and sometimes species composition (e.g., Flory and Milner, 2000; Biggs and Smith, 2002). These latter forms of true, nonfacilitative succession may represent various blends of the *tolerance and inhibition* successional categories described by Connell and Slatyer (1977).

Changes in types of succession and simple species replacement among and within FPZs may also be related to spatial differences in the degree and predictability of environmental fluctuations, including recurring seasonal patterns vs stochastic disturbances. We suspect that the more predictable the environmental changes in the riverine landscape (i.e., less stochasticity), the less likely that true succession dominates. Or another way of stating this hypothesis is that, "simple species replacement increases its importance over true succession (facilitative and nonfacilitative) as hydrological variability decreases and predictability increases in the wetted portion of the riverscape." Even here, the degree of change should depend on where the habitat is located along gradients of connectivity in the riverscape or even between the riverscape and the floodscape. The explanation for this perhaps counterintuitive hypothesis is that species can adapt to, or be selected for, a habitat whose seasonal conditions are predictable (even if demanding). In some cases, major components of the community are maintained throughout the year, while in other instances, there is a somewhat predictable replacement of species as seasons progress. The latter should be especially true for algae. In contrast, when hydrologic patterns become more variable and less predictable (two related but different concepts), then habitat conditions are altered more significantly in space and/or time, competitive ranks change, and the patterns of species replacement and community composition are disrupted. This promotes true succession in the riverine landscape vs the former pattern of simple species replacement with seasons, and the ultimate nature of the community will depend on species life history adaptations to the new conditions. Eventually, of course, the ecosystem adjusts to the new hydrologic pattern of variability and predictability (assuming it does not change again), and species replacement may return to its prior importance.

Although most successional studies in the riverine landscape have focused at the macro-scale, the importance and form of succession could be different in the microscale landscapes of the meiobenthos (e.g., Robertson and Milner, 1999) and lotic biofilms (e.g., Burns and Walker, 2000; Jackson *et al.*, 2001). The latter includes temporal changes in richness, density, and relative abundance of bacterial, algal, and other components of the biofilm in response to some combination of autogenic processes (including potential resource competition) and allogenic (seasonal) changes in the environment (Lyautey *et al.*, 2005). Bacteria seem to dominate in early stages of biofilm succession (Sheldon and Walker, 1997) followed by cyanobacteria and later eukaryotic algae, with the latter dependent on light penetration (Burns and Walker, 2000). It is not clear at present what forms of succession dominate biofilm communities, but this process does *not* appear limited to simple species replacement. Changes in biofilm characteristics can affect grazers and have been implicated in the extinction of at least one group: snails in the lower Murray River (Sheldon and Walker, 1997).

Little is known about how biofilm succession varies along longitudinal, lateral, and to some extent vertical dimensions in a riverine ecosystem except to the extent that ratios of bacteria, fungi, and algae should be affected by the amount of light limitation from canopy cover, turbidity, depth, and substrate. Much of the biofilm work has involved leaf litter and rock surfaces, and these substrates clearly vary by FPZ and longitudinal position in a riverine ecosystem. Where large amounts of wood are present in a riverine ecosystem, such as in forested headwaters and FPZs with complex geomorphology and forested islands (often in medium to large rivers), biofilms should also be important. However, we are unaware of studies showing differences in basic successional patterns for rock and wood substrates, which would portend in themselves any longitudinal differences in basic types of biofilm succession.

Proper management, conservation, and rehabilitation of rivers require an understanding of how predictable changes in flow regimes along with artificially generated stochasticity will alter successional processes in different ways from headwaters to large rivers. For example, a shift in flood frequency or predictability could change the competitive balance on the floodplain between pioneer and canopy species and between native and exotic species (e.g., Deiller *et al.*, 2001), both of which would alter the nature of the floodscape and ultimately the organic and inorganic inputs into the riverscape.

ECOSYSTEM AND RIVERINE LANDSCAPE PROCESSES

Model Tenet 10: Primary Productivity Within Functional Process Zones

Primary productivity in riverine landscapes varies with the type of functional process zone, based on their inherent environmental characteristics, which primarily include hydrologic retention and connectivity, geomorphic complexity, the light environment, and inputs from the riparian zone and watershed.

Clinal perspectives on longitudinal patterns in primary productivity (e.g., the RCC; Vannote et al., 1980) initially emphasized changes in the light environment of the main channel. Autotrophic production in headwaters ranges from low to high, depending on whether the canopy cover is extensive to minimal, respectively. As the channel widens and deepens downstream, the potential canopy cover from the riparian zone typically decreases. Consequently, benthic autotrophy by bryophytes, vascular macrophytes, and various attached microalgae rises in mid-order streams in forested watersheds (but not necessarily in prairie streams; Huggins, Thorp, and Baker, unpublished data). Net primary production was thought to peak in mid-order streams, the only site where P/R ratios were thought to regularly exceed 1. Benthic autotrophic production per square meter of the channel then declines downstream in the main channel with a drop in the percentage of the channel bottom receiving photosynthetically active radiation (PAR). [Autotrophic production near the banks should remain high, and this is the site of maximum benthic invertebrate diversity (Thorp, 1992). Inorganic turbidity also affects the availability of PAR, but the relationship between turbidity and stream size is not always consistent (Huggins Thorp, and Baker, unpublished data).] Gross phytoplankton production is usually considered to be important only in large rivers, but phytoplankton may be more crucial to food webs of small streams than previously thought, especially in grassland ecoregions.

A significant deficiency in the earlier and common linear perspective of rivers[1] as it related to autotrophic production is that the lateral components of the riverine landscape were ignored. This was partially resolved for the floodscape – but not the riverscape – by the publication of the FPC by Junk et al. (1989) and a concurrent modification of the RCC by Sedell et al. (1989). Within the lateral components of the riverscape in geomorphically complex FPZs are a great diversity of suitable habitats for unicellular and multicellular autotrophs, many taxa of which do not tolerate more than minimal water currents. The relationship between stream size and total primary production for the entire riverscape (not just the main channel and the floodscape) has not been examined empirically, as far as we know. It would be instructive to test whether net production is greatest in mid-order systems, as previously thought, or instead peaks in large-order reaches of rivers with extensive floodscapes.

Although a broad-scale, downstream pattern of primary production in the main channel should be present in all rivers, to understand patterns in autotrophic production at any point in the riverine landscape, one needs information on the nature of the FPZ in that area. Autotrophy should be greatest for FPZs characterized by higher lateral geomorphic complexity, more hydrologic retention zones, and larger areas of substrate exposed to PAR. Higher lateral complexity increases the amount of shallow water habitats as well as overall hydrologic retentiveness. Combined, these characteristics provide more habitats for submerged, floating, and emergent macrophytes (each colonized by attached algae) along with abundant assemblages of phytoplankton. System productivity will also vary among FPZs according to hydrologic connectivity within the riverscape (positive and negative attributes of current flow, including replenishment of inorganic nutrients), bed characteristics, and riparian–riverscape interactions affecting inputs from the watershed.

Model Tenet 11: Riverscape Food Web Pathways

On an average annual basis, autochthonous autotrophy provides, through an algal–grazer food web pathway, the trophic basis for most metazoan productivity for the riverscape as a whole, but allochthonous organic matter may be more important for some species and seasons and in

---

[1] See section entitled "THE LATERAL DIMENSION OF RIVERS – THE RIVERINE LANDSCAPE" in Chapter 2 for an historical discussion of the linear portrayals of rivers.

shallow, heavily canopied headwaters; however, a collateral and weakly linked, decomposer food pathway (the microbial-viral loop) is primarily responsible (with algal respiration in some cases) for a river's heterotrophic state ($P/R < 1$).

Until the publication of the RES, no riverine model other than the RCC (Vannote *et al.*, 1980) and its modifications (e.g., Ward and Stanford, 1983b; Minshall *et al.*, 1985; Sedell *et al.*, 1989) had addressed changes in food resources or functional feeding groups along the entire riverine ecosystem. The RCC stipulated for a generalized river that the primary food sources along the longitudinal dimension of the riverine ecosystem are the following: (i) allochthonous organic matter (principally riparian leaves) in headwater streams with a heavy canopy cover; (ii) benthic algae, mosses, and vascular macrophytes in shallow, low- to mid-order rivers; and (iii) FPOM in large rivers that was derived from terrestrial organic matter, which had "leaked" from upstream food webs. The RCC's predictions for large rivers were revised (Sedell *et al.*, 1989) to incorporate conclusions of the FPC (Junk *et al.*, 1989; Junk and Wantzen, 2004) on the contributions of the floodscape to food webs of large floodplain rivers. The RES, the RCC, and the FPC share some characteristics, but significant differences occur in their perspectives on food webs in the riverscape and the floodscape, especially as they pertain to large rivers.

The primary source of energy supporting metazoan food webs in headwaters seem to be a combination of autochthonous production, riparian inputs of particulate and dissolved organic matter, and terrestrial invertebrates, with the relative balance dependent on the extent of canopy cover (Vannote *et al.*, 1980). Using stable isotope analysis, Finlay (2001) showed that consumer $\delta^{13}C$ values[2] in temperate headwaters through medium-sized rivers are more strongly related to algal than terrestrial carbon. However, he found a clear transition from terrestrial to algal carbon sources as watershed area increased, a parameter that is indirectly associated with a decrease in percentage of the river covered by a riparian canopy. In headwaters of semiarid and arid ecoregions, where canopy cover is rarely extensive, autochthonous autotrophy should be the dominant source of energy for the food web. This is especially significant from a global perspective because a glance at a biome map of the world will show that more than half of the terrestrial ecoregions in tropical and temperate zones are arid or semiarid. [It would be instructive to know how past riverine theories would have differed if their authors had worked primarily in nonforested ecoregions!]

Riverine ecologists generally agree that autochthonous autotrophy is usually the dominant energy source for Metazoa in mid-order streams/rivers because such systems are typically shallow with minimal canopy and adequate light for benthic microalgae (Vannote *et al.*, 1980). Mosses and vascular macrophytes can also be relatively abundant in mid-order streams but are unlikely to contribute in a substantial and direct way to production of most Metazoa (cf. Winemiller, 2004) and are probably not even a major indirect contributor because of their recalcitrant detritus. However, their decomposition may significantly augment system metabolism in some rivers through microbial processing.

The source of energy fueling large river food webs continues to be more controversial than in smaller rivers and streams. A popular view based on the FPC (Junk *et al.*, 1989) and revised RCC (e.g., Sedell *et al.*, 1989) has been that organic matter exported from the floodscape to the riverscape is the principal nutrient source in floodplain rivers rather than FPOM leaking from upstream (the initial proposal of the RCC). This organic matter is thought to be derived from decaying terrestrial detritus in submerged floodplains and aquatic macrophytes growing in floodscape lakes. This perspective was apparently based on assumed ecological responses to

---

[2] The abbreviation $\delta^{13}C$ is derived from the ratio of ^{13}C to ^{12}C (compared to the $\delta^{13}C$ of a standard); for more information on the use of stable isotopes, see Fry (2006).

the spatially extensive and highly predictable Amazonian flood pulses and then extended conceptually to temperate-zone rivers. Unfortunately, the extrapolation was based on minimal empirical evidence – a common malady plaguing many aquatic theories. The ecological premise was that the huge amount of decaying terrestrial organic matter on the floodplain combined with growth of vascular macrophytes in floodscape lakes would support the production of Metazoa and other system components throughout the riverine landscape. Eventually, however, two problems in the trophic component[3] of the FPC and revised RCC became apparent when specifying energy sources for communities in the riverscape of large rivers. First, flood pulses in most temperate floodplain rivers are considerably more abbreviated and less predictable than those in tropical rainforest rivers. Consequently, trophic adaptation to seasonal flood pulses would be considerably more problematic for species living in rivers of the temperate zone (cf. Winemiller, 2004). Thus, it would seem to be more selective for communities in a temperate riverscape to be based on reliable autochthonous sources of organic matter, such as pelagic or benthic microalgae (Thorp and Delong, 1994). Second, very little evidence has come to light demonstrating that organic matter generated on the floodscape, especially the less recalcitrant material, is actually exported to the riverscape in large amounts. Indeed, the opposite seems to be the case. In channels of the Orinoco River of Venezuela, one of the world's largest rivers, nonfloodplain sources accounted annually for 63% of transported phytoplanktonic carbon because floodplains conserved algal production and acted as closed systems (Lewis, 1988; Lewis *et al.*, 1990). Although a substantial portion of POC (particulate organic carbon) in some tropical rivers is apparently derived from catchment vegetation (e.g., Bird *et al.*, 1998), this carbon may have originated primarily from tributaries, as little carbon, nitrogen, and phosphorus leak from tropical floodplains to the main channel (Junk, 1980 in Junk *et al.*, 1989; Hamilton and Lewis, 1987). This carbon should also be much more recalcitrant than autochthonous carbon produced in the riverscape.

To summarize, organic matter derived from the floodscape as a result of a flood pulse should rarely be a reliable energy source for at least temperate-zone rivers because export itself is apparently minimal from most floodplain rivers, and the flood pulse for temperate river is not sufficiently long or predictable to provide a dependable and sufficiently long-lasting source of energy for species living in the riverscape, even if they migrate for short periods into the floodscape. If this conclusion is true, then either FPOM leaking from upstream (as originally proposed by the RCC) or autotrophy within the riverscape (Thorp and Delong, 1994, 2002) must provide the major energy source within the riverscape of most large rivers.

Although food web studies remain scarce in riverine ecosystems (as do most ecological studies), a growing body of research conducted in tropical through subarctic rivers in moist and arid environments on several continents suggests that autochthonous primary production is an important and often majority contributor to metazoan production in rivers. The relatively extensive, empirical evidence supporting the importance of autochthonous production in large river food webs and the minimal, conflicting evidence were critiqued in Thorp and Delong (2002). For example, in a review of their ~15 years of research on the Orinoco and its flood-plain, Lewis *et al.* (2001) identified phytoplankton and periphyton as the ultimate organic matter source for most invertebrates and fish, even though macrophytes and terrestrial litterfall composed 98% of potentially available carbon. Their conclusions for a tropical river were largely based on stable isotope and production data. We have drawn similar conclusions for the north temperate Missouri, Ohio, and Upper Mississippi Rivers using stable isotope data (Thorp *et al.*, 1998; Delong *et al.*, 2001; Delong and Thorp, 2006). Since the 2002 review, the evidence

---

[3] The scientifically important FPC consists of a very persuasive hydrologic component (the ecological role of physical flood pulses) and a more controversial trophic model.

supporting the dominant role of autochthonous carbon in river food webs has continued to grow (e.g., Bunn *et al.*, 2003; Wilczek *et al.*, 2005; Cotner *et al.*, 2006).

Why should upstream-produced FPOM be less important to metazoan food webs in large rivers, given that large amounts of POM and DOM are known to be in suspension? The primary explanation may be that FPOM from upstream has undergone multiple metabolic and biochemical transformations and is typically quite recalcitrant by the time it reaches the large river food web, as acknowledged by the RCC. [However, some photochemical processes can increase DOM (dissolved organic matter) lability as the organic matter passes downstream (Opsahl and Benner, 1998 for the Mississippi River).] Therefore, its value as a nutritive source is far less than the relatively labile, benthic and pelagic microalgae (Thorp and Delong, 1994). Moreover, the long-held assumption in many riverine models that most suspended FPOM in large rivers is derived from terrestrial carbon (originally CPOM) leaked from upstream food webs may, in fact, be incorrect, at least during some seasons. Recent chemical and microscopical analyses of seston composition have demonstrated that living and detrital autochthonous matter, principally phytoplankton, is an important or even a major constituent of transported POM in temperate and some tropical rivers (Delong and Thorp, 2006). This has been shown for the Mississippi, Colorado, Rio Grande, and Columbia Rivers (Kendall *et al.*, 2001), for a floodplain reach of the River Danube (Hein *et al.*, 2003), and for the Upper Mississippi's main channel during the summer (Delong and Thorp, 2006). Although seston composition varies among seasons and rivers, the predominance of autochthonous organic matter in seston extends for much of the year but is especially prevalent during periods of maximum consumer productivity. [The composition and origin of benthic organic matter, however, is relatively unknown and may be different from suspended material.]

The major challenge to conclusions on the importance of autochthonous production has been that most riverine ecosystems are thought to be microbially based, heterotrophic ecosystems (e.g., Cole and Caraco, 2001) in which heterotrophic utilization of organic matter exceeds autochthonous autotrophic production (Wetzel, 2001, p. 733). This observation resulted in the *heterotrophy paradox* (Thorp and Delong, 2002): "How can animal biomass within riverine food webs be fueled primarily by autochthonous autotrophic production if the ecosystem as a whole is heterotrophic?" The response to this paradox, as discussed in Thorp and Delong (2002), is essentially that autochthonous autotrophy, through an algal–grazer pathway, provides the majority of energy for metazoan metabolism, but that respiration of dead organic matter from combined allochthonous and autochthonous sources shifts the balance of whole system metabolism on an annual basis into a heterotrophic state. A decomposer (microbial–viral loop) food pathway may process much of the transported, allochthonous and autochthonous carbon and contributes substantially to the heterotrophic state ($P/R < 1$) of many large rivers while only contributing in a minor way to metazoan production (Thorp and Delong, 2002). A weak coupling between microbial and metazoan production has also been noted by Lewis *et al.* (2001) and Delong and Thorp (2006).

We conclude, therefore, that autochthonous autotrophy provides >50% of the mean annual energy supporting metazoan community production in the riverscape for riverine ecosystems as a whole (from headwaters to the mouth). This applies to *all* types of rivers, not just to constricted channel systems. [This contrasts with limitations specified in Thorp and Delong (1994) but is in agreement with conclusions of Thorp and Delong (2002).] Although the metazoan community as a whole depends primarily on instream primary production, individual species may rely on allochthonous organic matter for most of their energy annually and/or seasonally. Allochthonous production may provide maintenance energy for many species during lean times of the year even though the same taxa may depend on autochthonous sources for growth and reproduction during periods of greater autotrophic production.

This food web trend identified above for riverine ecosystems in general will vary in degree among individual types of FPZs. Exceptions will occur, for example, where autotrophs have not adequately adapted photosynthetically to low light or nutrients caused by factors such as a dense, enclosing canopy in headwaters, certain geological features, and possibly high inorganic turbidity. This restriction is consistent with conclusions of the RCC for forested headwaters. One might expect the greatest reliance on microalgae over allochthonous carbon in hydrogeomorphically complex FPZs, such as the Upper Mississippi River, where hydrologic retention areas are more prevalent and autochthonous production is correspondingly greater. In contrast, the relative contributions of autochthonous autotrophy should diminish in FPZs with constricted channels, such as the tidal freshwater Hudson River. However, even in these systems, autochthonous production may represent over 50% of the energy for metazoan production – as demonstrated for the constricted-channel Ohio River (Thorp and Delong, 2002). The trophic basis of production should shift from a relative emphasis on benthic to pelagic autotrophs from headwaters to mouth but will also be influenced by the nature of the local FPZ.

Geomorphically complex FPZs are characterized by a greater amount of shoreline per longitudinal river kilometer than found in FPZs with fewer channels, and this has a number of trophic implications for terrestrial inputs and exchanges. For example, the increase in riparian area per linear river mile will enhance terrestrial–aquatic links by increasing inputs of terrestrial insects to aquatic food webs. This in turn could influence aquatic community composition by both increasing the number of fishes that eat invertebrates at the water surface and, as a consequence, lessening predation pressure on aquatic invertebrates (Nakano *et al.*, 1999). The latter could increase invertebrate competitive interactions if densities of benthic invertebrates rise. From the opposite perspective, the increased shoreline could enhance densities and perhaps richness of terrestrial predators (e.g., spiders) and scavengers feeding on prey that originated in the aquatic environment (Arscott *et al.*, 2005).

Model Tenet 12: Floodscape Food Web Pathways

> Algal production is the primary source of organic energy fueling aquatic metazoan food webs in floodscapes of most riverine landscapes, especially in rivers with seasonal, warm-weather floods; the relative importance of autochthonous autotrophy in floodscapes decreases with a rise in the amount and temporal extent of inorganic turbid conditions.

On the basis of earlier studies by Bayley (1989), Junk *et al.* (1989), and others, most aquatic ecologists probably still accept the proposition that metazoan food webs in flooded riverine landscapes are primarily fueled by direct or detrital consumption of vascular macrophytes, decomposing terrestrial plant matter, and occasional seeds, fruits (primarily in the tropics), and terrestrial invertebrates. However, this hypothesis on the importance of allochthonous carbon to Metazoa as a whole has been challenged for tropical and temperate floodplain rivers (e.g., Hamilton *et al.*, 1992; Forsberg *et al.*, 1993; Lewis *et al.*, 2001; Winemiller, 2004). These studies support the preeminent role in metazoan food webs of grazer and detrital consumption of microalgae growing at the bottom, attached to macrophytes, or suspended in the water column of submerged floodscape habitats. This may seem anomalous because of the inherently larger biomass of terrestrial and aquatic macrophyte detritus in floodscapes, but apparently much of the FPOM assimilated by detritivorous fish is derived from algae, even in floodplains where macrophytes dominate primary production (Araujo-Lima *et al.*, 1986; Winemiller, 2004; Winemiller *et al.*, 2007). Moreover, fish production in most rivers is dominated overall by relatively few species and short food chains, even in species-rich tropical floodplain rivers (Winemiller, 2004). This efficient transfer of energy may explain why large fish stocks in tropical floodplains can be supported by a seemingly minor component of the ecosystem's autotrophic production (Lewis *et al.*, 2001; Winemiller, 2004). This is important because the trophic transfer efficiency from decaying terrestrial vegetation and aquatic macrophytes is low as a result of either their recalcitrant nature or nonpalatable constituents (especially in tropical rain forests) (Thorp and Delong, 2002). Clearly, however, some floodplain ecosystems contain taxa specializing on the consumption of terrestrial plant matter and aquatic macrophytes, especially in the tropics (Winemiller, 2004). Such specialization would be difficult in most floodplain rivers in the temperate zone because of more abbreviated and less predictable flood pulses. The large amount of detrital material not consumed by aquatic Metazoa is decomposed by microbes in the river or ocean, buried by sediments, or remains on the floodplains and enters the terrestrial food web when waters recede.

It would be surprising if all aquatic food webs in a submerged floodscape behaved in the same fashion, and yet we tend to analyze them in this way or at least debate the concept in this manner. Should we expect the relative importance of decaying vegetation, vascular macrophytes, phytoplankton, and attached algae to be the same for submerged terrestrial floodplains and reconnected water bodies in the floodscape (isolated anabranches, oxbows/billabongs, and wetlands)? Autotrophic production, composition of the autotrophic assemblage, and relative abundance of suspended and attached algae all vary laterally in the riverscapes of some floodplain rivers, such as the Upper Mississippi River (Delong, unpublished data), and the same generic observation should apply to submerged floodscapes. Differences in species composition (e.g., suspended cyanobacteria vs green algae) among habitats would dramatically alter transfer of energy to Metazoa because of differences in palatability/toxicity, energy content, and omega-3 fatty acids. Moreover, different herbivores would be favored by changes in the relative abundance of attached and suspended microalgae aside from the species of autotrophs. Differences in vascular macrophyte species, such as submerged, floating, and emergent taxa, should also influence grazer food webs.

Another conceptual pitfall to avoid is the assumption that the same habitat in different kinds of floodscapes on different continents will behave in a similar fashion. Important differences in food webs within the same general type of habitat (such as submerged floodplains and reconnected floodscape lakes and wetlands) should be evident for floodscapes that have long-lasting, predictable flood pulses, more-abbreviated and less-predictable floods, floods separated by major droughts lasting over a year, and flood pulses occurring within or outside of periods of maximum potential rates of algal productivity (cf. Winemiller, 2004). Characteristics of floodscape soils and vegetation, which vary among FPZs, could also influence aquatic food webs in submerged floodscapes. For example, small particles on Australian floodplains stay suspended during a flood for long periods, whereas the inorganic particles settle relatively fast in floodscapes of the Amazon and the north temperate zone (Robertson *et al.*, 2001; Reid and Ogden, 2006). Increases in the amount and temporal extent of inorganic turbidity would cause declines in the amount of algal production, and thus increase the importance of allochthonous carbon to food webs in submerged floodscapes.

Model Tenet 13: Nutrient Spiraling

> Average nutrient spiraling length varies significantly among functional process zones, with spiral length: (i) increasing in a linear fashion directly with current velocity but changing abruptly at thresholds where boundary sediments are entrained; and (ii) decreasing with greater hydrologic retention and/or floodscape storage resulting from ·enhanced vertical and lateral geomorphic complexity.

Nutrient spiraling is a foundation concept of modern riverine ecosystem ecology, which combines nutrient cycling and downstream transport of solutes and gases (Webster and Patten, 1979; Newbold *et al.*, 1982; Newbold, 1992). Research has focused on biotic processing and physical retention by instream material (Webster, 2007), such as large woody debris (Valett et al., 2002) and the hyporheic zone. Despite the inherent longitudinal component of this concept, researchers have concentrated almost exclusively on spiraling in the main channel of headwater systems, undoubtedly because spiraling is easier to measure and manipulate in very small permanent and intermittent streams, such as with addition of ^{15}N to streams as a tracer (e.g., Mulholland et al., 2002). Although this research is clearly important, the potentially vital roles of lateral nutrient cycling and physical retention of nitrogen and other elements in medium to great rivers are greatly understudied.

Lateral habitats of the riverine landscape compress nutrient spirals in three basic ways. First, they retard downstream transport of solutes and gases because of minimal currents and floodscape storage of detritus and sediments. Second, biotic sequestration and biotic emigration are enhanced by: (i) greater primary and secondary production compared to main channel habitats; (ii) the presence of some longer-lived species compared to upstream (e.g., larger and older fish); and (iii) emigration to other ecosystems with waterfowl, which have been fed within slackwaters and alluvial wetlands. Third, anoxic microhabitats should be relatively more common in floodscape water bodies and in riverscape slackwaters than in the main channel and headwaters, thus affording more opportunities for microbially linked processes, such as nitrification and denitrification. Because the extent of lateral habitats varies among types of hydrogeomorphic patches, nutrient spiraling will differ significantly among categories of FPZs. These observations and conclusions call into question some important tenets of modern nutrient spiraling theory. As shown in Fig. 6.2, predictions on spiraling lengths can be significantly modified and even reversed depending on whether the focus is on the main channel, full riverscape, or entire riverine landscape (including floodplain storage).

Other environmental characteristics of FPZs will also influence rates of nutrient spiraling. Nitrogen processing in general is linked with the hydrologic cycle and its variability, including the extent to which threshold velocities are exceeded (Baldwin et al., 2005). Ratios of carbon to nitrogen should differ laterally within an FPZ and from one type of FPZ to another, and this C:N ratio has a known association with nutrient transformations, such as nitrification rates (e.g., Starry et al., 2005). Finally, geomorphically complex FPZs have a greater ratio of shoreline

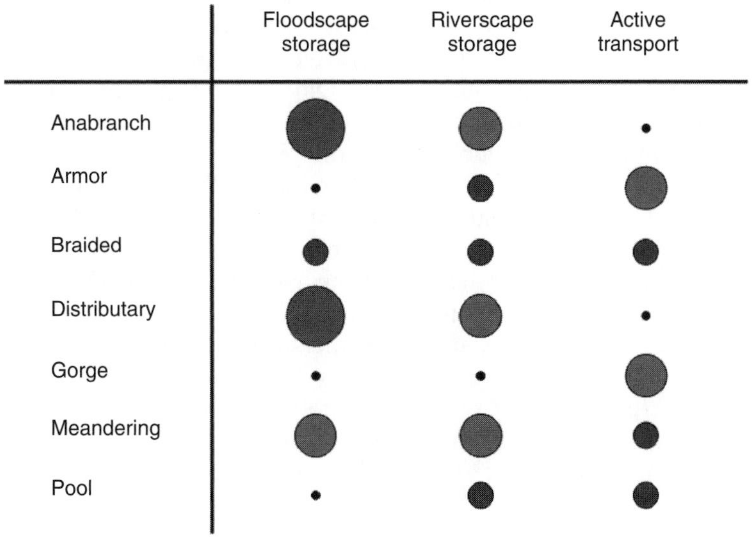

FIGURE 6.2   Potential for nutrient storage in the floodscape and the riverscape and active downstream transport in representative FPZs. Size of circles represents the relative amount of nutrient retention and active transport. This can be used to define the length of nutrient spirals for FPZs in Fig. 1.1 (See color plate 8).

length to longitudinal stream length and thus usually more riparia per river kilometer. The riparia reduce nutrient concentration into groundwater and the adjacent stream, affecting nutrient recharge (cf. O'Donnell and Jones, 2006).

Model Tenet 14: Dynamic Hydrology

> Naturally dynamic hydrological patterns are necessary to maintain the evolved biocomplexity in riverine ecosystems.

The importance of hydrology and hydraulics to community structure and ecosystem functioning has long been acknowledged by riverine ecologists,[4] most of whom have worked primarily in wadeable streams (e.g., Statzner and Higler, 1986), but it was not until the publication of the FPC (Junk *et al.*, 1989) that the role of floods as necessary ecological events in large rivers became widely acknowledged. The highly influential *natural flow regime* publication by Poff *et al.* (1997) not only demonstrated the crucial part played by floods in maintaining ecological integrity of naturally flowing rivers but also underscored their significance for river conservation, management, and rehabilitation. The five flow regime components often considered most important are discharge magnitude, frequency of a given level, duration of specific flow events, seasonality and predictability of flow events, and rate of change (flashiness) between levels. Seasonal changes in hydrology are especially important in the tropics where photoradiation and temperature show little seasonal variation.

Most of the conceptual and empirical attention paid to effects of changes in flow has focused on flood pulses. Floods are typically viewed as catastrophes and are thus strongly on the minds of the public, communication media, insurance companies, and governments. These flow episodes affect the floodscape (see Chapter 2) and are much more dramatic, albeit far less frequent, than flow pulses. Nonetheless, flow pulses may alter geomorphic complexity and other habitat characteristics in the riverscape that continually influences community structure and ecosystem processes. For example, as waters rise from base flow in the meandering sand bed Kansas River, the channel complexity varies in an inverse but nonlinear fashion with stream flow as the sand bars and secondary channels appear and disappear (see Fig. 6.2). During these transitions, current velocity rises, the bed moves, water quality changes (e.g., oxygen and turbidity), nutrient cycling is impacted as the aerobic/anaerobic environment of sandbars and the channel bed shifts, and the availability of hydrologic retention areas is altered (Fig. 6.3). These factors then influence habitat suitability for larval fish (e.g. Moore and Thorp, 2008), benthic invertebrates (O'Neill and Thorp, unpublished data), and both benthic and pelagic microalgae (Schmidt and Thorp, unpublished data).

Droughts are at the opposite hydrologic extreme from flood and flow pulses. These are almost entirely ignored by river ecologists except in arid and some semiarid regions of the world and are rarely studied by ecologists investigating wadeable streams except for researchers interested in the role of flow intermittency. Fortunately, the ecological role of droughts has recently begun receiving increased scientific scrutiny (e.g., Lake, 2003; Dodds *et al.*, 2004; Bonada *et al.*, 2006). A significant question is whether droughts operate as a greater mechanism of disturbance than do floods in riverine ecosystems (Sparks *et al.*, 1990; Delong *et al.*, 2001). From the opposite perspective, how long would it take for the natural or artificial *absence* of droughts to alter the species composition and natural ecosystem functioning in riverine ecosystems of arid to semiarid environment? Global climate change may be the factor providing the

---

[4] See section entitled "TEMPORAL DIMENSION: NORMALITY OR ABERRATION" in Chapter 2 for more historical information on the flow regime and additional details.

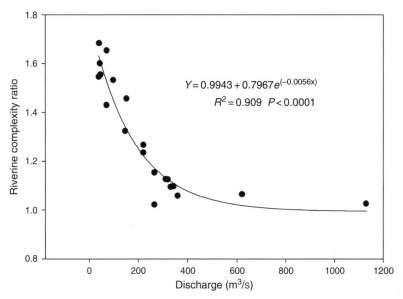

FIGURE 6.3   Changes in riverine complexity with discharge in the Kansas River. The riverine complexity ratio is the total length of exposed surfaces vs the total length of bankfull channel. Figure from O'Neill and Thorp (in review).

answer to this latter question because flow intermittency in headwater streams is predicted to increase in some semiarid environments, such as the U.S. Great Plains (Dodds *et al.*, 2004). New molecular techniques are also contributing insights to drought responses in microbial communities (e.g., Rees *et al.*, 2006).

Model Tenet 15: Flood-Linked Evolution

> The frequency of flood-linked life history characteristics increases directly with the seasonal predictability of floods and their concurrence with periods of maximum system primary productivity.

The importance of the flood pulse to life history characteristics of aquatic organisms and general biocomplexity was elegantly described for tropical rivers in the FPC (Junk *et al.*, 1989; Junk and Wantzen, 2004), but its role in temperate rivers has remained problematic. Recently, however, Winemiller (2004) explicitly linked the expression of flood-linked life history characteristics to the degree to which flooding occurs in phase with warm temperatures and enhanced system productivity. He identified three types of floodplain rivers: temperate stochastic, temperate seasonal, and tropical seasonal. In tropical rivers, reproduction/development, secondary production, and food web complexity are linked with highly predictable, long periods of flooding occurring against a backdrop of minimal seasonal fluctuations in temperature. In temperate stochastic rivers, linkage of life history characteristics with flooding is weakest with their aseasonal (unpredictable) flood regimes. Even in temperate seasonal floodplain rivers, maximum secondary production and selection for flood-linked life history characteristics depend on the flood pulse occurring in the late spring or summer.

   Species density and diversity in floodscape lakes respond directly to flood frequency as a result of adaptations of individual taxa to a variable aquatic environment. For example, the highest species richness of amphibians in the floodscape of the upper Rhone River occurred in habitat types with the greatest temporal variability (Joly and Morand, 1994). Some apparent adaptations to floods may, however, represent secondary adaptations in which the primary evolution was in response to another life need. For example, giant water bugs (Hemiptera: Belostomatidae) use rainfall as a cue of imminent flash flooding in small streams. Rather than seeking to exploit the floodscape, however, these bugs use the hydrologic cue to flee. This behavior may have developed as a dispersal mechanism to reach seasonal rain floods perhaps 150 myr ago (Lytle, 2002).

Model Tenet 16: Connectivity

> Biocomplexity generally peaks at intermediate levels of connectivity between the main channel and lateral aquatic habitats of the riverine landscape, but the relationship varies substantially among types of connectivity, evolutionary adaptation of taxa to flowing water, and functional processes examined.

Connectivity as a term is increasingly employed in fundamental and applied studies of rivers on various topics, such as patterns of species diversity for floodscape lakes connected at various stage height and submergence frequencies to conservation of species (e.g., Miranda, 2005). Spatiotemporal complexity of the riverine landscape – which varies among types of hydrogeomorphic patches – begets dynamic metastability and influences ecological biocomplexity. In riverine ecosystems, connectivity describes changes in processes and/or patterns over longitudinal, lateral, and vertical dimensions of a stream and usually implies an important temporal component as well (see review by Amoros and Bornette, 2002). Along a longitudinal dimension, connectivity describes the degree of surface and/or subsurface links among riffles and/or pools, especially under drought conditions (Bonada *et al.*, 2006; Marshall *et al.*, 2006; Sheldon and Thoms, 2006a) or in environments with abundant upwelling and downwelling areas, such as gravel bed rivers and karst spring systems (Dahm *et al.*, 2003). The vertical dimension of connectivity describes the amount of exchange between epigean waters and the hyporheic and, to a much lesser extent, phreatic zones. This will vary directly with the local climate and porosity of the bed and subsurface environments directly below and adjacent to the stream.

Across a lateral dimension, two principal forms of surface-water connectivity exist: riverscape and floodscape connectivity. The response of biocomplexity to these two forms of connectivity should be distinctive, but research on riverscape connectivity has lagged far behind studies of floodscape connectivity, possibly because of the latter's association with the FPC and the link to managed wetlands and economic losses from floods. Riverscape connectivity is also harder to measure than floodscape connectivity because the former is a continuous gradient while the latter is basically a present–absent phenomenon.

Riverscape connectivity is characterized by nearly continuous aquatic links at sub-bankfull river stages, with exchange rates ranging from high (adjacent to the thalweg and a flow-through environment) to very low (e.g., true backwaters linked at the surface only by flow pulses). Connectivity can be defined for the riverscape by the distance or time required for water to flow between the main channel and various slackwater habitats. Low connectivity implies that an area of the riverscape has high hydrologic retentiveness and older *water age*. Spatiotemporal aspects of riverscape connectivity are linked to both the present discharge level and the hydrologic regime. Increases in discharge, even within sub-bankfull river stages, can enhance exchanges between the channel and slackwaters (Delong and Bruesewitz, in review). Increased hydrologic connectivity should influence landscape diversity (i.e., structural complexity) in ways that will vary among rivers (Ward and Tockner, 2001), within rivers (Malard *et al.*, 2006), and among FPZs in particular. Also affected will be current velocities and, as a result, various organisms and processes affected by flow. These include, for example, oxygen tension, dissolved nutrient concentrations, organic content of the sediment, water temperatures, turbidity, productivity rates, community composition, FCL (Roach, Thorp, and Delong, unpublished data) dispersal strategies, and possibly the role of deterministic vs stochastic processes. Increased connectivity within the riverscape at high water should depress population densities, productivity, and community diversity for most aquatic macrophytes and for many Metazoa, algae, and prokaryotes that poorly tolerate currents, turbidity, and/or sedimentation. For some taxa, however, this negative relationship may be counterbalanced by advantages linked to currents, such as higher oxygen tension and greater additions of organic and inorganic nutrients to replace levels depressed by, for example, competition or sedimentation.

Floodscape connectivity features a disjointed pattern wherein floodplain water bodies (e.g., wetlands, lakes, and detached oxbows and anabranches) are severed from the riverscape for much of the year or even for multiple years. Floodscape connectivity is defined by the frequency of surface-water connection at supra-bankfull floods between the riverscape and various floodscape water bodies linked at different river stage heights. When the flood pulse brings river waters directly onto the floodscape, these water bodies change from temporally lentic systems to quasi-lotic habitats where water-driven advection may occur for short to long periods depending on the nature of the flood pulse. Waters from the river transport sediments, nutrients, and organisms into the floodscape lakes, altering directly or indirectly community composition and numerous ecosystem processes. The predictability, frequency, and length of the flood pulse may modify the relative importance of deterministic and stochastic processes in affecting community composition within the floodscape. Flood waters impact terrestrial vegetation and initiate or reset successional processes. Likewise, many floodscape lentic systems may follow a successional pathway from a fluvial origin to terrestrialization if not reset by subsequent floods.

The term connectivity is used extensively and somewhat loosely in the aquatic literature for riverine ecosystems with moderate to extensive floodscapes, but conceptual analyses linked to this potentially important process are rarely addressed and have not advanced very far (but see Amoros and Bornette, 2002). Biota, inorganic nutrients, and both POM and dissolved organic matter move at different rates among elements in the riverine landscape (cf. Wiens, 2002) according to spatiotemporal dimensions, adjacent patch context, abundance of edges, extent of corridors, and steepness of the boundary ecotones (which obviously differ for an organism vs a nutrient molecule). The basis for understanding this connectivity in riverine landscapes is mostly theoretical at this stage, and investigations have been limited primarily to hydrologic connectivity of the wetted riverscape with terrestrial and aquatic components of the floodscape (e.g., Bornette *et al.*, 1998; Tockner *et al.*, 1999a; Ward *et al.*, 2002; Hein *et al.*, 2003). Ward and Tockner (2001) concluded that overall biodiversity peaks at intermediate levels of connectivity, and Tockner *et al.* (1998) showed that the connectivity–biodiversity relationship varied considerably with the taxa being analyzed. Such fundamental research has also been applied directly to rehabilitation strategies (e.g., Hein *et al.*, 2005).

Although this model tenet predicts maximum biocomplexity at intermediate levels of connectivity, the current evidence supporting this seemingly simple relationship is sparse and the theoretical basis is not refined, especially when relying solely on the application of the IDH of Connell (1978) without a solid conceptual justification for riverine ecosystems. Plots of community structure or functional processes versus connectivity that result in intermediate relationships could produce not only a normal-shaped curve (as often illustrated for the IDH) but also ones characterized by extreme positive through negative skewness, as illustrated in Tockner *et al.* (1998). This could reflect differences in adaptation of taxa to aquatic environments in general (e.g., fish, amphibians, and birds) and to flowing waters in particular and the differences in functional processes examined. Clearly, additional research is needed in this area.

Model Tenet 17: Landscape Patterns of Functional Process Zones

Biocomplexity in riverine ecosystems is significantly influenced by the diversity, spatial arrangement, and character of the functional process zones.

Riverine landscape ecology is a relatively new and still small discipline within aquatic ecology, though it is based on the well-established field of (terrestrial) landscape ecology, which dates back to at least half a century (Forman and Godron, 1986). A very minor percentage of lotic ecologists would call themselves riverine landscape ecologists or even focus their research on the central themes of landscape ecology, which are applicable to rivers (see review in Wiens, 2002). However, many ecosystem ecologists in particular are working on peripheral components of this field, especially those with interests in effects of watersheds on streams, biotic and abiotic flows between the riverscape and the floodscape, and conservation biology along river corridors.

The focus of tenet 17 and the discussion in this chapter are on differences in riverine landscape ecology that are related to FPZs. The similarities and differences between terrestrial landscape ecology and riverine landscape ecology were analyzed by Wiens (2002), and thus will be touched on here only briefly.

The RES is fundamentally a landscape concept (see Fig. 1.2) because of its ecological emphasis on: (i) multidimensional aspects of riverine ecosystems; (ii) variations in hydrogeomorphic complexity ($\sim$ matrix heterogeneity) of riverine landscape elements with the nature of the local FPZ; and (iii) HPD across various spatial and temporal scales. Elements of this riverine landscape have been discussed previously but include aquatic and terrestrial components in both the riverscape and the floodscape along with surface and subsurface areas. Exchange rates of organisms and both inorganic and organic matter differ substantially among elements of the riverine landscape depending on whether the temporally dynamic boundaries crossed are terrestrial floodscape to aquatic floodscape or surface stream to either hyporheic, terrestrial riverscape, terrestrial floodscape, or aquatic floodscape. Exchanges also occur within the aquatic riverscapes at rates varying with the degree of connectivity (see tenet 16). Boundary permeability ranges from low to high according to the material or the organism moving across these patch edges or ecotones. Boundaries may be more important in riverine than terrestrial landscapes because they effectively intercept water-mediated flows and impede or trap transported materials and organisms (Palmer *et al.*, 2000; Wiens, 2002). They also represent an unusually severe ecotone because the transition is between aquatic to terrestrial elements of the landscape rather than, for example, transitions between forest and grassland. In comparison to terrestrial landscapes (Wiens, 2002), riverine landscapes are highly dynamic with a regulating, directional water flow that is variable in strength along four dimensions of the river.

Riverine landscape ecology can be studied at many hierarchical levels, but knowledge of the local FPZ and possibly of the adjacent FPZs should be critical in many cases. Major hydrogeomorphic differences among FPZs imply that patch quality will differ among FPZs, as well at smaller spatial scales. Patches representing various terrestrial and aquatic elements of the riverine landscape are mobile to differing degrees and may change shape and distribution. Over time, patches may also disappear and then reappear in the same or different location, especially smaller ones and those in the riverscape, at rates directly related to the hydrogeomorphic character of the FPZ. Boundary permeability will also change over time within an FPZ and at varying rates among FPZs according to the geomorphic structure (e.g., constricted vs braided channels) and hydrologic regime (e.g., Latterell *et al.*, 2006). For example, flow and flood pulses are likely to enhance boundary permeability by small to large degrees in constricted, meandering, and floodplain channel areas (in that order). This has been shown at smaller spatial scales for stream invertebrates (Lancaster, 2000). Processes operating within patches, such as nutrient cycling (Fisher *et al.*, 1998) and primary productivity, can be altered by patch configuration and changes in permeability with adjacent patches or elements of the landscape.

At the largest scale just encompassing the entire riverine ecosystem, FPZs represent primary patches along the longitudinal dimension. As such it is possible that their diversity and spatial configuration could be important to community composition (cf. Malard *et al.*, 2006), although in many cases this would be statistically and logistically difficult to demonstrate. In terms of community composition, the importance of these attributes would vary greatly with the motility of the organism. For instance, fish species that are habitat generalists may occur in multiple types of FPZs. However, the presence of certain types of FPZs, such as those with sizeable waterfalls, will influence fish distribution, especially if the barriers are located downstream. As an example, the presence of Glens Falls in the lower Hudson River, USA, acts as a barrier to upstream migration and dispersal of many fish species; indeed, the whole character of the river changes at this point (Jackson *et al.*, 2005). Species richness in a riverine ecosystem should be directly related to the diversity of FPZs because this would imply increased hydrogeomorphic complexity on a river-wide scale (see also tenet 3). It is not as clear whether the spatial configuration of FPZs would alter river-wide ecosystem processes, but the diversity of FPZs should have an impact. One might expect, for instance, overall primary productivity and rates of nutrient cycling to be increased in rivers with many FPZs compared to rivers with relatively few changes in channel structure, especially for those featuring relatively invariant channel structure, such as rivers with mostly constricted and simple meandering channels.

7

Ecogeomorphology of Altered Riverine Landscapes: Implications for Biocomplexity Tenets

Introduction
Distribution of species
Community regulation
Ecosystem and riverine landscape processes

INTRODUCTION

Stream ecosystem theories have, with few exceptions, been developed based on expectations about the functional dynamics of pristine systems. Efforts have been made to modify some models by considering both natural differences in characteristics of riverine ecosystems (e.g., Sedell *et al.*, 1989) and expected responses to diversity and function based on human-induced changes (e.g., Ward and Stanford, 1983b). The serial discontinuity concept (Ward and Stanford, 1983b), for example, proposed that dams function as a reset mechanism in riverine ecosystems, where fauna present below a dam are reflective of species either upstream or downstream of the dam. In this situation, the basis of the faunal change is based on the location along the longitudinal continuum of the riverine ecosystem. Existing theories have been used to predict responses to anthropogenic changes in riverine ecosystems, but explicit hypotheses on ecosystem behavior in altered landscapes have not been incorporated into these models. The world today necessitates the need for theory to have application to altered rivers, given that relatively pristine conditions are rare. Even rivers commonly viewed as pristine, such as the Amazon and Mackenzie rivers, do not escape the influence of humans from atmospheric pollution, global climate change, and the pressure to regulate discharge.

This chapter describes the types of responses expected in altered riverine ecosystems as defined by the tenets of the RES. While reading this chapter, keep in mind that anthropogenic disturbances do not exclusively affect only one of the domains (hydrology, geomorphology, and ecology) of the riverine landscape. Disturbances changing attributes in one domain will alter attributes of others through secondary or tertiary mechanisms (Fig. 7.1). No attempt is made to address every possible type of anthropogenic impact in altered systems. Instead, we address broad changes in the nature of riverine ecosystems through direct actions on attributes and give examples of deviations from expected patterns (Table 7.1). For example, how might one expect the nature of fish or invertebrate community composition to be altered below a hydroelectric dam? Structural and functional properties in the riverine landscape below a dam will depend on the nature of the FPZ present before the construction of the dam and the nature of the FPZ above and below the dam once it begins operation. Such an approach has been implemented successfully in assessing changes in FPZs as a result of flow diversions and water withdrawals in

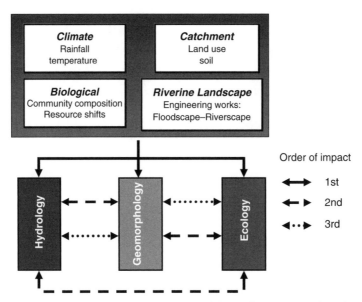

FIGURE 7.1 Anthropogenic disturbances and the immediacy of their influence on the three domains of the riverine landscape. Order of impact reflects if the effects are: (a) direct influence on a domain (first order); (b) indirect on the second domain through effects on an initial domain (second order); or (c) indirect through changes first imparted on two domains before influencing the third (third order). (See color plate 9).

TABLE 7.1 Examples of Changes in the Hydrological, Geomorphological, and Ecological Attributes of the Riverine Landscape Resulting from Anthropogenic Disturbances

Domain	Attribute	Examples
Hydrology		
	Magnitude	Reduction in peak flows, increased base flows, reduction or increase in mean annual flow
	Frequency	Increased occurrence of low-flow events, decrease in occurrence of smaller flow and flood pulses, increase in occurrence of mid-level flow events
	Duration	Prolonged low flows, reduction in length of overbank flows
	Timing	Seasonal reversal of flow events
	Rate of change	Increased rate of rise and fall of flow and flood pulses, increased stability of water levels
	Variability	Increases or decreases in flow events
	Predictability	Increased low-flow events, decreased events larger than flow pulses
	Patch character	Loss of abundance of hydraulic patches, redistribution of hydraulic patches
Geomorphology		
Riverine landscape	Connectivity	Increases and decreases in the flow and sediment loads
Floodscape	Stability	Increased and decreased erosion of floodplain surfaces
	Connectivity	Decreased flow connections between floodscape and riverscape, increased groundwater connections
	Dimensions	Reduction in the active floodscape area
	Patch character	Simplification and fragmentation of floodscape
	Composition	Loss of functional sets and units associated with floodscapes
Riverscape	Dimensions	Increased and decreased capacity of the active riverscape, change in shape and complexity of riverscape
	Planform	Straightening of the river channel, reduction in the number of river channels

(*continued*)

TABLE 7.1 (*continued*)

Domain	Attribute	Examples
	Slope	Downcutting of river channel
	Bed	Aggradation and degradation of the river bed, armoring of river bed sediment
	Patch character	Loss of functional sets and units within the riverscape, change in composition of functional sets and functional units
	Stability	Increased and decreased stability of the riverscape
Ecology		
	Competition	Introduction of nonnative species and changes in community composition
	Food resources	Shifts in importance or benthic or pelagic resources, altered primary productivity, changes in catchment and riparian inputs, changes in food quality
	Predation	Introduction of nonnative species, changes in community competition
	Reproduction and dispersal	Loss of reproductive and nursery habitats, disruption of migratory pathways

the Murray–Darling system (Thoms and Parsons, 2003). With this in mind, we again present the tenets of the RES by considering how they can be used to predict changes/responses from population to landscape levels. It is important to note that the tenets themselves do not change with alteration in natural conditions. Instead, the tenets provide a framework for applying the RES in situations encountered in altered systems and for understanding how changes in the characteristics of an FPZ influence ecosystem structure and function independent of location along a longitudinal gradient of a riverine ecosystem.

A goal of this chapter is to provide key policy makers and river managers with better tools for environmental management and rehabilitation of riverine landscapes. These tools could be employed, for example, in helping determine where to locate or remove dams, how far to set back channel levees, or in the development of assessment plans for rehabilitation projects. They could also help establish appropriate spatiotemporal linkages of main channels with artificially isolated, alluvial wetlands in the riverscape and the floodscape and could contribute to policies on seasonal water release patterns from reservoirs.

DISTRIBUTION OF SPECIES

Model Tenet 1: Hydrogeomorphic Patches

Species distributions in riverine ecosystems are associated primarily with the distribution of small to large spatial patches formed principally by hydrogeomorphic forces and modified by climate and vegetation.

Each FPZ possesses its own unique habitat template formed by the interaction between hydrological patches, which exist at small to large spatial scales, and the prevailing geology of the region in which the FPZ resides. Critical to the distribution and diversity of species within an FPZ is the number, diversity, and connectivity of patches that comprise the FPZ. Any changes, therefore, in the number, diversity, or connectivity of patches within an FPZ will influence the distribution and diversity of species occupying that FPZ or the FPZ that is created by alteration of the original FPZ.

Levees eliminate connectivity between the riverscape and the floodscape with an associated loss in lateral habitats of the riverscape and the floodscape (Fig. 7.2). The loss of conduits between these components of the riverine landscape disrupts life cycles of many organisms, particularly fish that rely on multiple patches to provide areas for foraging, reproduction,

FIGURE 7.2   Levee separating the riverscape and a narrow band of forested floodscape from floodscape used for agriculture along the lower Mississippi River, Mayersville, Mississippi, USA. Photo from the U.S. Geological Survey photo gallery (http://www.umesc.usgs.gov/aerial_photos/j/ja1h1_1997ob.html).

nurseries, and winter habitat in more temperate climates (Fremling *et al.*, 1989). Loss of both lateral and longitudinal linkages of patches has implications on the metapopulation dynamics of species, reducing their distribution by cutting off sources of recruitment. Such may be the case in the lower Illinois River where backwaters that are periodically connected to the river contain a greater number of species of freshwater mussels than those cut off from the river by levees that prevent fish hosts from carrying glochidia into some areas (Tucker *et al.*, 1996).

As a major feature in defining their nature, changes in flow characteristics will also influence the distribution of patches in an FPZ. Since the construction of the Glen Canyon Dam on the Colorado River, fine sediment accumulation has increased throughout the riverscape (Schmidt *et al.*, 2002). Additionally, reduction of the magnitude of snowmelt-driven floods diminished the scouring potential and the ability of the river to rework coarse-grained sediments. This lead to reduction or loss of return-current eddies, which serve as important fish nurseries and other habitats important for forage and spawning areas (e.g., Osmundson *et al.*, 2002). Loss of the ability of a river to sculpt the riverine landscape, therefore, leads to a reduction in the types of patches present and eliminates access to some patches, thereby changing the internal structure of an FPZ. Sedimentation from increased soil erosion from urbanization and agriculture creates a more homogeneous condition by filling interstices and reducing variability within patches. The result is a loss of habitat and an associated change in the distribution of species that require patches, which have been lost (Stewart *et al.*, 2005).

Model Tenet 2: Importance of Functional Process Zone Over Clinal Position

> Community diversity and the distributions of species and ecotypes from headwaters to a river's mouth primarily reflect the nature of the functional process zone rather than a clinal position along the longitudinal dimension of the riverine ecosystem.

An attribute common among many anthropogenic changes to riverine ecosystems is that physical characteristics are transformed to the point where the system looks and functions differently from its natural state. Alterations of a riverine ecosystem of sufficient scale can result in reduction of diversity of FPZs, transformation from one type of FPZ to another, or formation of FPZs unique to the network. Modifications in the number, arrangement, and types of FPZs result from changes in geomorphic and/or hydrological attributes, as these are the primary features defining FPZs

FIGURE 7.3   Channelized and culverted section of Dadu River, China. Photograph by Martin Thoms.

(see tenet 1 in Chapter 6). Consider the effects of channelization (Fig. 7.3). Depending on the length of river or stream that is channelized, several different FPZs can be changed into a single type of FPZ characterized by limited lateral complexity and altered hydrological retention through the homogenization of a sequence of FPZs (if channelization is extensive). Perhaps even more profound is the loss of longitudinal complexity resulting from reduction in the diversity of FPZs. This is illustrated by the Missouri River in the central part of the USA. Until early in the 20th century, the lower Missouri was characterized as a meandering, turbid river laden with islands (Funk and Robinson, 1974). Channelization changed the river below Sioux City, Iowa, into a narrow and swiftly flowing river, shortening the channel by 125 km and reducing the wetted area of the riverine landscape by nearly 64% (Whitley and Campbell, 1974). Annual and interannual hydrological variability of this region was further changed by the construction of six reservoirs in the middle reaches of this large river (Hesse and Mestl, 1993). Despite the channel's pathway through two physiographic provinces with different glacial histories and the effects of numerous large tributaries, the lower Missouri River now appears to function as a single FPZ where multiple FPZs probably existed (Galat *et al.*, 1998; Pegg and Pierce, 2002). As a result of many substantial environmental changes to the river, overall species diversity has declined and numerous species of fish, mammals, birds, invertebrates, reptiles, and plants are now listed as threatened or endangered (Galat *et al.*, 1998).

Both high, reservoir-forming dams (Fig. 7.4) and low-head navigation dams change the nature of FPZs, although the extent of impact will vary with the size and operation of the dams. The Upper Mississippi River is regulated for navigation by 26 low-head navigation dams (Fig. 7.5). These dams are not flood-control structures; instead, they maintain summer minimum flows at levels higher than historical conditions to maintain a 2.7-m deep channel for commercial navigation (Delong, 2005). The navigation dam system on the Upper Mississippi creates a series of navigation reaches (commonly labeled with the misleading term "pool"), with each navigation reach possessing conditions close to the predam environment in the upper two-thirds of the reach. The lower one-third of a navigation reach, however, is markedly different from predam conditions. Water levels in this lower area of the navigation reach are continually high as a result of proximity to the lock and dam, creating hydrological characteristics very different from the upper part of the reach (Sparks, 1995). The result of this design is that while the upper two-thirds of each navigation reach reflect the predam FPZ, the lower

FIGURE 7.4   Katse Dam, a water storage dam on the Malibamatso River, Kingdom of Lesotho. Photograph by Martin Thoms.

FIGURE 7.5   Lock and Navigation Dam 5A and spillway, Winona, MN, USA. Photo from the U.S. Geological Survey photo gallery (http://www.umesc.usgs.gov/aerial_photos/e/eu3a6_1997ob.html).

one-third of each navigation reach represents a new FPZ. This different FPZ possesses an entirely new set of hydrological and geomorphological parameters that redefine species distribution and community structure longitudinally in the Upper Mississippi River. The FPZs present prior to the construction of the dams are now a series of repeatable pockets separated by new, repeating FPZs (the lower one-third of each navigation reach). Species diversity and community composition have consequently changed through reductions in the distribution and diversity of aquatic vegetation in lower part of each navigation reach, as well as loss of some riverine benthic invertebrates (Delong, 2005). We can estimate the effects of these changes by comparing fish assemblages in two rivers whose FPZs were once similar, the relatively unaltered lower Wisconsin River and the Upper Mississippi River. Limnophilic species have become increasingly common in the lower, impounded portions of the Upper Mississippi, while a more riverine fauna has been retained in the Wisconsin River (Weigel *et al.*, 2006).

The impact of high dams built for flood control and hydroelectrical production on aquatic and terrestrial species composition has been well established (e.g., Petts, 1984). The serial discontinuity concept addressed this issue in the context of a clinal perspective of rivers, stating that dams shift community structure below a dam to resemble communities either upstream *or* downstream of the dam (Ward and Stanford, 1983b). The serial discontinuity concept predicts that return to expected conditions and species composition is dependent on the point on the longitudinal gradient of a riverine ecosystem where the dam is placed. Studies testing the concept, however, have suggested that channel geomorphology, hydrological variability, and lateral connectivity also strongly influence the recovery process (Stevens *et al.*, 1997; Casas *et al.*, 2000; Jungwirth *et al.*, 2000). Given that these features are also important in characterizing an FPZ, changes in species composition above and below a dam will reflect the nature of the FPZ prior to and following impoundment. Clearly, the impounded section will represent a new FPZ, given the newly permanently wetted areas and changes in hydrological attributes. Species composition below the impoundment, rather than reflecting communities either upstream or downstream of the impoundment, will resemble areas within the riverine ecosystem with a comparable FPZ or an entirely new FPZ.

Changes in erosional processes and nutrient inputs from the floodscape and basin landscape also have the potential to blur distinctive attributes of FPZs. Lapwai Creek, Idaho, USA, flows through three distinct geomorphological forms over its 48-km length. It is likely that these regions represent three different FPZs, especially when hydrological changes are also considered. Fish and invertebrate community structure remains quite similar along the length of this agriculturally impacted stream despite these obvious differences in the landscape and stream attributes (Delong and Brusven, 1998), suggesting that the distinctive community structure expected along the length of the stream has been lost as nutrient additions and sedimentation bring a constancy to the structural and functional processes of the stream.

Model Tenet 3: Ecological Nodes

Species diversity is maximum at or near ecological nodes representing hydrogeomorphic transitions between and within functional process zones and in areas of marked habitat convergence (e.g., tributaries) and divergence.

As described in Chapter 6, borders denote areas of overlap between hydrogeomorphic patches at any spatial scale within hierarchical framework of the RES. As such, geomorphic and/or hydrological transitional zones are areas where a broader range of conditions exist, which are suitable for habitation by a greater array of species. Ecological nodes will serve the same purpose in altered systems. The question becomes, however, how are the distribution, frequency of occurrence, and types of nodes influenced by deviations from the natural condition? In most cases, existing confluences of tributaries with the main channel remain after the river has been altered. The rare exceptions occur where rivers are diverted into different basins. Perhaps more significant from an ecological perspective is the loss of geomorphic transitions in areas where levees or reservoirs have changed the geomorphic structure of the bed and lateral areas or the water's depth and flow velocity. The ecological effects of deviations from the natural presence of tributary and geomorphic nodes will vary with the extent to which nodes are lost, diminished in function, or newly formed.

Issues addressed in the previous two tenets certainly have implications on the presence of nodes. Channelization and the construction of levees diminish patch abundance, thereby reducing the number of nodes present through alteration of hydrological characteristics and the loss of lateral and longitudinal complexity (Galat *et al.*, 1998; Schiemer *et al.*, 2001a). Similarly, dams may contribute to loss of nodes or formation of different types of nodes. Sand or gravel bars and islands would be points of convergence and divergence impacted by all these disturbances. As critical physical attributes of riverine ecosystems of all sizes, islands and bars expand patch availability for invertebrates, fish, terrestrial vertebrates, and aquatic vegetation with implications to abundance, diversity, and reproductive success (Nichols *et al.*, 1989; Thorp, 1992; Kessler, 1998; Desgranges and Jobin, 2003; Hurley *et al.*, 2004; Thorp and Mantovani, 2005). Loss of islands through diminished capacity for moving and transporting materials for island bed formation or loss through inundation eliminates major transitions associated directly with an island and reduces habitat associated with islands. These reduce the availability of snags and nearshore retention zones, which are important to the diversity and abundance of aquatic organisms for the entire riverscape.

Tributaries commonly represent important transition areas where diversity and species abundance are higher relative to adjacent patches (Minshall *et al.*, 1985). Linkages to tributaries can be changed through actions occurring within the tributary basin, such as nutrient enrichment, increased soil erosion, or water removal, as well as through other anthropogenic disturbances already discussed. The nature of the node, however, is changed if the tributary enters at a point where the other stream is impounded. If there is considerable disparity in size between two merging streams, impoundment effects are likely to extend upstream into the tributary, changing the location of the transition or eliminating it altogether. Fish species migrating upstream or downstream in response to environmental conditions or reproductive state may not only find pathways blocked but also encounter new competitors and dangerous predators. For example, many small species of fish in prairie streams migrating downstream because of droughts now face a gauntlet of voracious largemouth bass (*Micropterus salmoides*); this is thought to have contributed to the endangerment of federally listed species such as the Topeka shiner (*Notropis topeka*) (Winston *et al.*, 1991; Knight and Gido, 2005).

With the presence of transitional areas dependent on the type and location of patches within an FPZ, any process leading to a homogenization of patches may eliminate some ecological nodes (e.g., Boyero and Bailey, 2001). Erosion, sedimentation, and nutrient enrichment, as major impacts from agricultural, forestry, and urban land use, serve as agents that may reduce patch heterogeneity and alter the arrangement of patches in riverine landscapes. These agents can operate separately or together to induce changes in the configuration and heterogeneity of hydrogeomorphic patches within an FPZ and associated changes in species diversity. Comparisons of five small Southland streams in New Zealand revealed that the combined effect of erosion from grazing and loss of riparian vegetation leads to a community dominated by benthic invertebrates tolerant of warm water and dense periphyton (Chironomidae and the caddis fly *Oxeythira albiceps*) compared to the taxa preferring cold water and low abundance of periphyton (including many Plecoptera and the caddis fly *Helicopsyche albescens*) (Quinn *et al.*, 1992).

Model Tenet 4: Hydrologic Retention

> Overall community complexity varies directly with the diversity of hydrologic habitats in a functional process zone and increases directly with hydrologic retention until other abiotic environmental conditions (e.g., oxygen, temperature, substrate type, and nutrient availability) become restrictive.

Hydrologically retentive zones, ranging from marginal slackwater habitats and eddies of streams to floodscape lakes, augment diversity and enhance productivity of phytoplankton, aquatic macrophytes, zooplankton, benthic invertebrates, and fish (Eckblad *et al.*, 1984; Hildrew, 1996; Bini *et al.*, 2001; Schiemer *et al.*, 2001a; Koel, 2004; Akbulut, 2005). Hydrological retention is also important for storage and processing of organic matter and nutrients (Aspetsberger *et al.*, 2002; van der Nat *et al.*, 2003; Hein *et al.*, 2005) and influences the role of abiotic and biotic factors in the regulation of community composition (Baranyi *et al.*, 2002;

Thorp and Mantovani, 2005). A key attribute of hydrological retention is, therefore, the duration of a retentive zone, which is influenced by the extent of hydrologic connectivity between patches within the riverine landscape. Consequently, the relationship between hydrological retention and community complexity will be directly influenced by changes in the arrangement and types of patches, hydrological processes that promote connectivity, geomorphological attributes, and water quality.

Actions that alter hydrological variability, patterns of flow and flood pulse, or connectivity between patches of high and low hydrological retention will have particularly strong impacts on community complexity. Either the reduction of hydrologic variability, as occurs in an impounded area, or marked increase in variability, as often seen in urban streams, could eliminate hydrological retention within an FPZ or at least alter its function and effectiveness. Hydropower peaking relies on the storage of water in a reservoir until increased electrical production is needed to meet greater demand. This creates a daily cycle of prolonged low-flow conditions with short intervals of high discharge. In the case of the Chippewa River, Wisconsin, outflow from the last dam on the river remained at $14 \, m^3/s$ for 16-20 h each day during the summer but was raised quickly to $142 \, m^3/s$ to satisfy increased daily electrical demand. Habitat suitability analysis for fish and invertebrate communities indicated that fauna present for more than 40 km below the dam were dominated by generalists with a limited occurrence of riverine specialists. This was largely due to the instability of hydrologically retentive areas (Delong and Mundahl, 1995; Low and Lyons, 1995). In contrast, water diversions increase hydrological retention and stability, but to the point of reducing or eliminating lateral and longitudinal linkages between patches. Water diversions in systems ranging from the Gila River, Arizona, to Cooper Creek, Australia, have caused substantial declines in native fish and invertebrate taxa and abundance (Sheldon *et al.*, 2002; Rinne *et al.*, 2005). Changes in the Gila River have actually promoted the success of nonnative species through declines in patch resources for native species and the creation of hydrogeomorphic patches more suitable for nonnatives.

Just as flow pulses are important in determining hydrological retention within the riverscape, flood pulses shape hydrological retention in the floodscape. Distance from the riverscape, elevation, and proximity to ephemeral channels on the floodplain have implications for water quality and community composition of floodplain lakes under natural conditions (Petry *et al.*, 2003; Arscott *et al.*, 2003). The entire floodscape is subjected to periodic physical restructuring and inflows of nutrients, sediments, and organic matter. Loss of connectivity decreases hydrological retention of the terrestrial and aquatic floodscape. The potential cost is the loss of periodic inputs that lead to the enhanced productivity characteristic of the floodscape and causes the loss of species dependent on temporal disruptions of hydrological retention and patterns of connectivity with the riverscape (Tockner *et al.*, 1999a).

Circumstances may arise where hydrological retention in both the riverscape and the floodscape appear to function normally, but community composition and diversity are different than expected. As storage sites of nutrients and organic matter, retentive zones are susceptible to inputs that alter water quality and lead to the occurrence of species tolerant of these conditions. Increased sedimentation can change the bathymetry of a channel, thereby altering flow patterns. Retentive areas used by species that can access these zones only by moving through stream segments with low to moderate flow may no longer be available if sediment deposits shift the thalweg toward the access point of the retention zone. Although not as subtle, removal of riparian and floodplain vegetation as well as snag removal operations leads to the loss of small, but important retention zones formed by large woody debris. The importance of large woody debris as sites of retention of nutrients and organic matter has been well-documented as has their role in habitat formation directly through their physical structure and indirectly through disruption of flow (Lehtinen *et al.*, 1997; van der Nat *et al.*, 2003; Tockner *et al.*, 2006).

COMMUNITY REGULATION

Model Tenet 5: Hierarchical Habitat Template

> The most important environmental feature responsible for ecological regulation of community composition is a hierarchical habitat template, as determined primarily by interactions between geomorphic habitat features and both short- and long-term flow characteristics.

The literature is ripe with examples of how anthropogenic perturbations affect the distribution and abundance of organisms in the riverine landscape. Numerous metrics based on abiotic conditions have been established to assess the biotic integrity or health of aquatic communities. Perhaps it is stating the obvious to say that the geomorphic and hydrologic components of the habitat template will be modified as a function of the type of environmental perturbation, and the direction and degree of this change will vary with the scale of the environmental perturbation. It is important, however, to emphasize within the context of this statement that the *scale* of change in geomorphic and hydrological features will be a reflection of the scale of the perturbation. This is probably the reason that some metrics used in developing indices of biotic integrity and comparable assessment tools fail to detect a biotic response to disturbance, even when it is obvious from the observation that some measurable deviation or difference should be present. It is not the methodologies that are flawed; rather, it is that metrics used to assess the impact are not measuring responses at the scale of the disturbance (e.g., Strayer *et al.*, 2000; Lepori *et al.*, 2005).

Metrics for assessing the effects of a localized chemical spill, for example, may be most appropriate if they operate at the scale of a functional unit or a functional channel set. Functional process zone-scaled measures would be best applied for assessing the impact of hydrological patterns or changes in land use (e.g., Lorenz *et al.*, 1997). Also to be considered, in addition to the scale of the perturbation, is the scale of the habitat template of the organism(s) of interest. Migratory fish, for example, move in response to seasonal habitat needs associated with reproduction, feeding, or overwintering (Fremling *et al.*, 1989). Failure to include appropriate metrics that define their habitat needs and the corridors needed to move between habitats (all of which are going to be forged from the hydrogeomorphic features of the system) must be considered in understanding the implications of anthropogenic changes to the habitat template of the biota (Leland and Porter, 2000; Aarts *et al.*, 2004).

Conceived under the context of the RCC as the most viable conceptual model of riverine biocomplexity present at the time, the serial discontinuity concept was restricted to a clinal perspective for characterizing how man-made disturbances alter the status of the abiotic factors that define habitat templates. Changes imparted by high dams for flood control and hydro-power, as an example, change water temperature patterns both daily and seasonally (Caissie, 2006), restrict movements between critical habitats (Gore and Bryant, 1986), alter substratum characteristics (Stevens *et al.*, 1997), and modify diel and seasonal conditions for water quality (Santucci *et al.*, 2005). What occurs, therefore, is a redistribution, loss, or formation of new hydrogeomorphic patches in the FPZ, where any deviation beyond the tolerance range of biota will reshape the habitat template sufficiently to exclude some species found prior to the disturbance, while establishing templates suitable for the establishment of biota not present in the original FPZ. All this ensues because most disturbances at the scale of an FPZ redefine a stream's hydrological and geomorphic characteristics (Poff *et al.*, 1997; Stevens *et al.*, 1997; Thoms and Parsons 2003), leading to reconfiguration of the habitat templates evident at smaller spatial scales within the impacted area. Rather than characterizing resets of the biota along a longitudinal gradient, the serial discontinuity concept characterizes how disturbances impacting large spatial scales create FPZs not originally present in a given location. These novel FPZs are representative of an FPZ(s) located elsewhere in the riverine ecosystem, which possess hydro-geomorphic attributes that generate similar habitat templates.

Model Tenet 6: Deterministic vs Stochastic Factors

Both deterministic and stochastic factors contribute significantly to ecological regulation of communities, but their relative importance is scale- and habitat-dependent and affected by the organism's lifespan vs timescale of environmental fluctuations; however, stochastic factors are more important overall and throughout the riverine ecosystem.

An important attribute of specific human impacts on riverine ecosystems is whether the disturbance decreases or increases stochasticity, thus altering community regulation. A simple explanation of this is to consider community-level responses to the frequency, magnitude, or timing of flood and flow pulses. Studies have demonstrated a direct correlation between hydrological retention time and species composition of zooplankton communities in rivers. Baranyi *et al.* (2002) noted that rotifers dominated the zooplankton community of the Danube River from low to moderate levels of hydrological retention. It was not until hydrological retention reached a sufficient duration that biotic controls regulated community composition. Rotifers were consistently the primary representatives of the zooplankton of prairie rivers in the central United States, which are characterized by pronounced natural hydrological variability, whereas crustacean zooplankton abundance was much greater in the more hydrologically stable Ohio and St. Lawrence rivers (Thorp and Mantovani, 2005). Contrast this to observations in three streams in northwestern Mississippi, USA, where hydrological variability has increased through a combination of channelization, watershed deforestation, and stream incision. Fish communities in these systems were characterized as *colonizing assemblages* (or r-selected populations) whose presence was attributed to increased stochasticity associated with greater flashiness in the hydrograph (Adams *et al.*, 2004).

Unfortunately, it is difficult to ascertain whether shifts in community composition in riverine ecosystems with artificially reduced hydrological variability are more attributable to a diminished role of stochastic factors or an increased role of deterministic factors. The potential for the persistence of nonnative fish is an excellent example of this conundrum. Nonnative fish species are common in sections of the Verde River, Arizona, where floodplain agriculture and water diversions have reduced hydrological variability, but nonnative species are nearly absent in sections of the river where changes in hydrological characteristics are minimal (Rinne, 2005). Valdez *et al.* (2001) proposed that managed floods in controlled rivers, such as the lower Colorado River, USA, have the potential to benefit native species and reduce the success of nonnative predators and competitors. In both these instances, it is likely that a reduction in the natural stochasticity diminishes the selective advantage of native species.

Another consideration in the relationship between stochastic vs deterministic controls is the distribution of patches within an FPZ. The examples used for zooplankton and the significance of hydrological retention reflect the presence of spaces within the river's geomorphological framework that promotes retention during low- to moderate-flow periods (*sensu* Schiemer *et al.*, 2001a). The type, diversity, and arrangement of patches present within an FPZ will be strongly influenced by the nature of hydrological conditions, with hydrological variability often influencing access to patches (Schlosser, 1987; Arrington and Winemiller, 2006). A critical theme here is that the diversity of available patches and their distribution, which are largely influenced by stochastic events in riverine ecosystems, can minimize the influence of competition and predation on community structure.

Any human-caused perturbation that creates a redistribution of patches, loss of patches, or the addition of patches different from those typically found within an FPZ has the capacity to change the mechanisms controlling community structure if the level of stability in environmental conditions changes significantly.

Model Tenet 7: Quasi-Equilibrium

A quasi-equilibrium is maintained by a dynamic patch mosaic and is scale-dependent.

Discussion of the previous tenet proposes that anthropogenic disturbances shift the balance or nature of stochastic and deterministic regulation of community structure in riverine ecosystems. The degree of change in the status of factors regulating community structure will dictate if a quasi-equilibrium can be maintained. In some natural communities, higher-species diversity depends on processes that interrupt competitive exclusion and allow competitively subordinate species to thrive. For example, occasional stream spates may temporarily favor a subordinate metazoan species and allow it to coexist over long periods with the competitive dominant because: (i) it better tolerates the higher flow conditions; (ii) it is more adept at recruitment to unexploited patches via the aquatic stage or by propagules; or (iii) the formerly limiting resource is no longer restrictive as a result of the stream spate reducing total consumer densities. If humans alter the diversity and variability of patch types (small to large scale) in space or time, then competition can reduce species diversity. Under these circumstances, any future equilibrial conditions are likely to occur with a much lower diversity of species. Identifying the presence of, or change in, a quasi-equilibrial state is dependent on the spatiotemporal scale of observation.

Ecologists earlier considered that community and ecosystem equilibrium were associated with lower environmental variability, but we now better appreciate that it is the disruption of *natural patterns*, whether characterized by low or high environmental variability, that shifts the system away from an equilibrial or quasi-equilibrial state. Discussion of model tenet 6 emphasized how reduction in hydrological variability would increase the significance of deterministic controls through reduced stochasticity. Commonly associated with this is a decrease in species adapted to use the myriad of hydrogeomorphic patches created by variable flow. This can result from a loss of habitat or species replacement by more generalist and nonnative species that often characterizes community structure in hydrologically disturbed environments. Although deterministic processes may become more important in shaping community structure through competition with and predation on native species over short temporal scales, the community may, in fact, become more susceptible to stochastic factors over the long term, with a concomitant reduction in opportunities for establishing and maintaining a quasi-equilibrial state. Although flood control efforts may, on average, diminish the magnitude and duration of flow and flood pulses, it is ironic that events overpowering our efforts to control floods, such as when levees are breached, can substantially increase the negative impact of large-scale events. The occurrence of these extreme stochastic events is likely to be sufficient to limit the potential for a system to return to a quasi-equilibrium. Conversely, anthropogenic activities that magnify hydrological variability increase the stochasticity of riverine ecosystems, probably to the point that conditions do not persist for sufficient duration to allow deterministic controls to contribute to community structure. Streams located in watersheds where changes in land use have led to increased frequency of flow pulses and short-duration flood pulses are often characterized by fish communities composed of *colonizing* species. These communities reflect the absence of periods of stable flow lasting long enough to allow establishment of species intolerant of highly variable flow conditions (Adams *et al.*, 2004). Long-term maintenance of these conditions is likely to diminish the contribution of deterministic controls at any scale and would prevent the system from achieving a quasi-equilibrium.

Land use and morphological changes in the riverine landscape and the drainage basin can reduce the potential to achieve a quasi-equilibrium. Such disturbances alter the dynamics of the patch mosaic both spatially and temporally. Deforestation (Fig. 7.6), urbanization (Fig. 7.7), agriculture, and other practices that change vegetation conditions alter runoff characteristics and increase erosion (Ziegler *et al.*, 2004; Wantzen, 2006). This leads to changes in the timing

FIGURE 7.6   Clear-cut along the Lower Cotter River, ACT, Australia. Photograph by Martin Thoms.

FIGURE 7.7   Aerial photograph of the Danube River at Vienna, Austria, highlighting changes in the riverine landscape from urbanization, including straightening of channel and disconnection of lateral patches. Image accessed through the NASA and the Applied Research and Technology Project Office (https://zulu.ssc.nasa.gov/mrsid/mrsid.pl). Final preparation of image by Carol Lowenberg.

and quantity of nutrients and sediments entering the system, altering system productivity and changing bedform characteristics (Gregory, 2006; Orr *et al.*, 2006). Removal of floodplain and riparian vegetation leads to greater diel and seasonal fluctuation in water temperature (Hetrick *et al.*, 1998; Dunham *et al.*, 2007), further compounding the influence of stochastic factors. These disturbances also have the potential to decrease patch diversity within an FPZ by converting once diverse patches into patches possessing similar profiles for physicochemical and autotrophic production (Delong and Brusven, 1992, 1998). This homogenization of patches within an FPZ would reduce the frequency of refugial patches that serve as sources for recolonization following a stochastic event (e.g., Townsend, 1989). Land use is not the only means by which the significance of stochastic regulation can be enhanced. Although typically associated with hydrological control, both low-head and high dams increase the daily and seasonal fluctuations of temperature, pH, sediment transport, and dissolved oxygen concentrations, adding a stochastic component to their influence (Santucci *et al.*, 2005). As described earlier, dams also change the distribution of patches within an FPZ and can create new FPZs (Gillette *et al.*, 2005; Tiemann *et al.*, 2005), thereby altering the dynamic interaction between patches and changing the conditions sustaining the quasi-equilibrium.

Model Tenet 8: Trophic Complexity

> Food chain length (maximum trophic position) increases directly with the geomorphic complexity of a functional process zone in response to multiple factors related to habitat heterogeneity, diversity and abundance of food resources, and dynamic stability.

Trophic complexity is a reflection of the types and numbers of species present, the apical predator present, and the quality and quantity of resources available. An emphasis of the RES is that the spatial and temporal characteristics of these factors are dictated by the hydrological and geomorphic complexity of an FPZ. This physical complexity, then, influences trophic complexity in general and trophic position in particular. This reflects the observation that increased geomorphic complexity provides a greater mosaic of patches, with these patches and their nodes providing space for a larger diversity of consumers and producers. The nature of the relationship, whether linear or curvilinear, remains to be seen, but the expected response should be that human alterations in the riverine landscape that reduce hydrological or geomorphic complexity will reduce trophic complexity.

The linkage between trophic dynamics and the attributes that define an FPZ has only recently been considered. Walters *et al.* (2007) observed that bed morphology influences the chemical characteristics of seston and detritus. Doyle (2006) developed a model that proposed changes in channel geomorphology and flow patterns can result in the exclusion of predators. This exclusion would benefit prey, with likely implications to the availability of basal resources. Hydrological conditions alone can influence trophic dynamics and the mechanisms controlling trophic complexity. Marks *et al.* (2000) examined food chain length during drought and flood years in the Eel River, California, USA. In their study, predator densities declined during drought years and grazers controlled algal growth. In contrast, predator densities increased during flood years with a corresponding increase in algal growth, suggesting a change from top-down controls during flood years to bottom-up controls during drought years.

All these examples suggest that the geomorphological complexity of an FPZ will influence trophic complexity through shifts in food resources or predator–prey interactions. Post and Takimoto (2007) proposed that food chain length can change by: (i) the addition or removal of intermediate predators or basal species (insertion mechanism); (ii) changes in the degree of trophic omnivory of the top predator (omnivory mechanism); and (iii) changes in the trophic position of intermediate predators that are prey for the top predators, which is caused by insertion and omnivory mechanisms at lower trophic levels. The examples previously described fit the proposed mechanisms through exclusion of predators, possibly even the top predator, or changes in resources available, which would have implications for the extent of omnivory at lower trophic levels. It should be expected, therefore, that human disturbances to an FPZ will elicit changes in trophic complexity and food web length by reducing geomorphic complexity. Changes in habitat template, as described in model tenet #5 in this chapter, reduce species diversity, including predators, alter the storage and transport of organic matter, and modify the potential for autochthonous production. These actions would initiate any of the three mechanisms proposed by Post and Takimoto (2007), thereby reducing food chain length and trophic complexity across the FPZ.

Lateral connectivity between riverscape and floodscape provides an example of how trophic complexity correlates with geomorphic complexity of an FPZ. River–floodplain ecosystems are characterized as possessing an array of lateral habitats within the riverscape plus aquatic and terrestrial habitats of the floodscape. The level of complexity of this array will depend on the nature of the specific FPZ (i.e., not all river–floodplain FPZs are created equally). Productivity of habitats within the riverine landscape will vary among FPZs, as will the resources available. Connectivity between these foraging patches offers consumers, especially highly mobile species, a broader range of resources. This increases the potential of omnivory beginning with primary consumers and continuing to the highest trophic level. This is exemplified by a study comparing the carbon, nitrogen, and sulfur stable isotope ratios of benthic-feeding fish in the lower Mississippi River, the Atchafalaya River basin, and the coastal floodplain of upper Barataria Bay in Louisiana, USA (Fry, 2002). Fish from the Mississippi River exhibited distinct isotopic ratios, which reflected the loss of access to the aquatic habitats of the floodscape. Levees along the lower Mississippi River have cut off access to resources important to not only the adult fish, but also those used by invertebrates and juvenile fish. Fry's study did not examine trophic complexity directly, but it did illustrate that access to a diversity of patches can influence energy flow.

Reduced lateral geomorphic complexity eliminates conduits between these patches and restricts the resources available. The likely outcome in this situation is a reduction in the potential for omnivory and concomitant shortening of food chain length. Calculation of trophic position of fish from the lower Ohio, Upper Mississippi, and lower Missouri rivers found that trophic position of all major fish feeding guilds was consistently lowest in the Missouri River (Fig. 7.8). Geomorphic complexity of the study areas of the lower Ohio and Upper Mississippi rivers largely remains, whereas lateral complexity and connectivity to the floodplain have been greatly reduced in the lower Missouri River by channelization and levees.

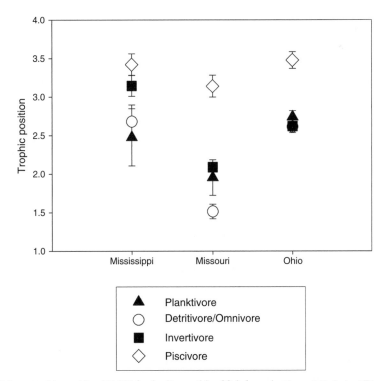

FIGURE 7.8   Mean trophic position (±1 SE) for feeding guilds of fish from the Upper Mississippi River near Louisiana, Missouri; lower Ohio River near Metropolis, Illinois; and the lower Missouri River near Washington, Missouri. Data used for calculating trophic positions were originally presented by Delong *et al.* (2001).

A decline in the potential for omnivory is not the only factor that contributes to reduced trophic complexity, nor are levees the only mechanism for reducing lateral connectivity. Inundation of the floodplain in tropical and many temperate rivers provides spawning habitat for a number of riverine fish. Juvenile fish produced on the floodplain, in turn, serve as a prey for large piscivores (Winemiller, 2005). Adult fish can reproduce, albeit with less success in suboptimal habitats in the absence of a flood pulse. The resulting drop in densities of juvenile fish reduces prey availability, thus requiring larger predators to find alternate prey, which may be lower on the food chain. Flood-control dams, water diversions, and other human disturbances that change the magnitude, timing, and duration of a flood pulse should, therefore, cause a shortening of the food chain through changes at intermediate levels of the food chain brought about by these reductions in hydrological complexity of an FPZ.

Geomorphic complexity is not limited to the lateral dimension of an FPZ. Smaller-scale patches within an FPZ form a longitudinal array, with the diversity and connectivity between these patches also serving to influence trophic complexity. Jepsen and Winemiller (2002) observed that trophic position was higher in fish from nutrient-poor blackwater rivers than from more productive rivers. A later study of the Cinaruco River, one of the low-productivity rivers examined by Jepsen and Winemiller (2002), discovered that bocachico, *Semaprochilodus kneri*, a major consumer of organic-rich sediment migrates from the productive Orinoco River to the Cinaruco during the low-water period. During this time, it is a major contributor to the biomass of piscivores in the Cinaruco, particularly *Cichla* spp. (Layman *et al.*, 2005; Winemiller *et al.*, 2006). The capacity to move between rivers, or patches in the case of an FPZ, and take advantage of more productive areas both laterally and longitudinally should be a critical component in the relationship between trophic and geomorphic complexity. Construction of dams, weirs, or even indirect changes in physical attributes of a riverine landscape would constrict or severe the migratory pathways of prey species, changing trophic complexity through the loss of an intraguild species.

Model Tenet 9: Succession

The relative importance of different types of succession (simple species replacement, true facilitative succession, and true no-facilitative succession) varies between the riverscape and the floodscape, among functional process zones, and across gradients of hydrologic variability. Classical (facilitative) succession is primarily limited to terrestrial elements of the riverscape (on ephemeral islands) and the floodscape (riparian and floodplain habitats), and it occurs in response to hydrogeomorphic processes. Simple species replacement in the wetted elements of the riverscape increases in relative importance over true succession (facilitative and non-facilitative) as hydrologic variability decreases.

Succession in the riverine landscape, as this tenet states, should be a function of the prevailing hydrogeomophic characteristics of an FPZ. The nature of any type of succession in disturbed networks will, therefore, be dictated by the hydrogeomorphic attributes present following alterations to the system. Principal among these will be the change in stochasticity wrought by how a disturbance influences hydrological variability and predictability. Specifically, changes in the riverine landscape that lead to greater hydrological variability will serve as a constant disruption in succession, causing continual resets of community composition in both the riverscape and the floodscape. The expected result would be a community dominated by colonizing, r-selected species adapted to harsh environments. The fish assemblages of streams where hydrological variability has increased as a result of channelization and deforestation are dominated by small-bodied, early maturing species, whereas larger-bodied species are rare even though they would typically be present under more normal hydrological conditions (e.g., Adams *et al.*, 2004). Zooplankton and benthic invertebrate communities would be expected to respond similarly, as would benthic invertebrates, given the linkage between hydrological variability and community structure of zooplankton (Baranyi *et al.*, 2002; Arscott *et al.*, 2005; Thorp and Mantovani, 2005). These studies found that the periodicity and duration of high flows restricted crustacean zooplankton, leading to a community dominated by rotifers, which can take advantage of increased hydrological variability as long as protective slackwater areas persist. Similarly, successional patterns of the flora and fauna of the floodscape and terrestrial components of the riverscape are continually reset by hydrological processes. Increased hydrological variability limits flora of the terrestrial components of the riverine landscape to those species adapted for using moist soils or can cope with short interflood periods (Jaromilek *et al.*, 2001), while excluding species where the interflood period is too short in duration to allow them to become sufficiently established to withstand periodic inundation.

Reduction in hydrological variability and loss of connectivity should elicit similar changes in patterns of succession as both represent a decline in the exposure to stochastic controls. Periodic inundation of the terrestrial floodscape is essential in the sculpting of the geomorphological features that influence species composition and distribution on the floodscape (Bendix and Hupp, 2000) as well as the successional changes on the floodscape (Cordes *et al.*, 1997; Karrenberg *et al.*, 2002), seed dynamics (Nilsson *et al.*, 1991), and the seed bank structure (Bornette *et al.*, 1998; Goodson *et al.*, 2000). The loss of periodic floodplain inundation consistent with the adaptation of flood-tolerant species could allow succession to proceed without interruption. Periodic inundation of the terrestrial floodscape along the Red Deer River, Alberta, Canada, was the major mechanism for replenishment of poplar stands (Cordes *et al.*, 1997). Flood regulation, however, is likely to eliminate this process, causing poplar stands to continue to senescence much like stands that were naturally removed from flooding by sedimentary terracing. Construction of dams and embankments as well as water removal has reduced the number of small- to medium-sized floods in the fifth-order Brenna River, Switzerland, allowing floodplain communities to advance to late successional stages (Brunke, 2002). Although these situations arise from eliminating floods as a means of eliminating mature trees from the floodscape and creating gaps for new vegetation growth, loss of connectivity from reducing hydrological variability or by physical obstruction (e.g., levees) eliminates a means of dispersing seeds and propagules of plants important to succession dynamics that are transported into the floodscape with rising waters (e.g., Jansson *et al.*, 2005). The outcome of the minimization or the elimination of hydrological variability would be that deterministic controls would become more important in shaping temporal changes in community composition in both the terrestrial and aquatic realms of the riverine landscape.

Linkages between geomorphic features and dynamic hydrology should not be ignored when developing riverine restoration plans. Natural geomorphic features of the floodscape, as Brunke (2002) pointed out, are an important determinant for successional dynamics. The same would be true in the aquatic realms of the riverine landscape, as these physical features will dictate the species present within a network (the implications of disturbance are described in this chapter under tenet 5), which will, in turn, set the parameters for species replacement. This will be true whether dealing with issues pertaining to channel or floodplain modification or changes in land use, because these and other alterations change the physical configuration of an FPZ, with implications at all scales. The importance of woody debris is one area where there are linkages between geomorphic properties of an FPZ and changes in successional patterns. In addition to serving as a substrate and refuge for biota of a riverine ecosystem, the interaction of flow and sedimentation with large woody debris leads to the formation of important morphological features of the riverscape and the floodscape (Tockner *et al.*, 2006). Creation of a new habitat template within an FPZ can occur in the absence of large woody debris (where stochastic controls help maintain pioneer communities) or where woody debris is overabundant and succession is allowed to continue to senescence.

ECOSYSTEM AND RIVERINE LANDSCAPE PROCESSES

Model Tenet 10: Primary Productivity Within Functional Process Zones

Primary productivity in riverine landscapes varies with the type of functional process zone, based on their inherent environmental characteristics, which primarily include hydrologic retention and connectivity, geomorphic complexity, the light environment, and inputs from the riparian zone and watershed.

As this tenet implies, primary productivity will be dependent on the type and distribution of patches within an FPZ. The physical requirements of bed characteristics, light, nutrients, plus patterns of flow and potential hydrologic retention for lotic autotrophs have been well studied, particularly at smaller spatial scales. What becomes important at larger scales, including the level of an FPZ, is how the patches are arranged with regard to access to light, nutrients, retentive zones, and hydrological attributes, the last of these being important with regard to delivery of nutrients from the riparian zone and drainage basin. Modifications of any of these attributes, therefore, will have ramifications for primary production in an FPZ.

Loss of physical heterogeneity in riverine landscapes has been an issue of much concern, with human activities leading to a homogenization of the physical template (Stanford *et al.*, 1995; Meyer, 1996). Sedimentation caused by changes in land use and bank destabilization leads to a reduction in bed heterogeneity by covering substrates and filling interstitial spaces (Hansmann and Phinney, 1973; Iwata *et al.*, 2003). An additional confounding effect of disturbances that lead to sediment deposition is that turbidity is increased and may remain a constraint on primary production even during low-flow periods. In this case, homogenization of patches is both spatial (uniformly high turbidity laterally and longitudinally in an FPZ) and temporal (shorter intervals between periods of high turbidity). Nutrient enrichment can create a more homogeneous environment for algal growth. The importance of retentive zones in the form of benthic structure or irregularities in morphology that increase hydrological retention are minimized if nutrients are widely distributed and available in a stream or river. Conversely, the removal of physical or hydrological retentive zones eliminates patches where nutrients may be more readily available (Naiman *et al.*, 1986; Shields and Smith, 1992; Hein *et al.*, 2003).

Hydrological processes function to shape geomorphic complexity as well as control for other physical conditions important to primary production (Poff *et al.*, 1997). Many of the strategies for managing flow are designed to attenuate flow and flood pulses, which reduce the capacity of a river or stream to alter its geomorphic landscape. Channelization creates a more uniform pattern of flow laterally and longitudinally within a channel. The higher flows by themselves reduce the potential for primary production of phytoplankton through the loss of hydrologically retentive zones. Loss of shoreline sinuosity shrinks the amount of surface area available for benthic production and causes an overall reduction in the geomorphic complexity of an FPZ. Both flood-control dams and water diversions serve to reduce hydrological variability. Although greater hydrological stability under these conditions will reduce overall complexity of the riverscape spatially and temporally, the effect will be to actually increase primary production, particularly for phytoplankton. Minimization of flood and flow pulses increases hydrological retention, creating a protracted period of growth and reproduction (Sellers and Bukaveckas, 2003). Retention would be increased still more above a flood-control dam with a general response of increased phytoplankton production and possible declines in benthic autochthonous production resulting from the formation of this new FPZ.

Another feature influencing productivity in an FPZ will be the presence and extent of interaction with a floodplain. In riverine landscapes, productivity in the terrestrial and aquatic components of the floodscape contributes substantially to productivity of the riverine landscape (Lewis *et al.*, 2000). Conversely, periodic inundation can serve as a determinant for the types of autotrophs and resource availability for the terrestrial and aquatic floodscape (Pouilly *et al.*, 2004; Schilling and Lockaby, 2006; Gruberts *et al.*, 2007). Reduction or elimination of periodic inundation of the floodscape by flood-control dams, levees, and water abstraction would eliminate this contribution to overall system productivity in complex FPZs.

Some undisturbed FPZs, because of low complexity, limited light penetration, low hydrologic retention, nutrient limitations, or some combinations of these, will exhibit low rates of primary productivity. Changes in the nature of any of these controls, however, could serve to diminish or enhance primary production within the FPZ. Nutrient changes resulting from the alteration of the floodscape or the terrestrial landscape within or above an FPZ will create a potential for primary productivity beyond that expected in an FPZ with limited complexity (Delong and Brusven, 1991; Kiffney *et al.*, 2003, 2004). Increased light penetration through complete or even partial removal of shading by riparian vegetation will increase primary production regardless of the complexity of an FPZ (Hetrick *et al.*, 1998; Kiffney *et al.*, 2004). This latter point summarizes the implications of human disturbance on primary productivity within an FPZ. Primary productivity will increase in concordance with geomorphic complexity due to greater heterogeneity (e.g., Cardinale *et al.*, 2002). Processes that reduce geomorphic complexity will often lead to a decrease in primary productivity unless other factors important to autochthonous production (light, nutrients, and hydrologic retention) compensate for naturally or anthropogenically low complexity.

Model Tenet 11: Riverscape Food Web Pathways

> On an average annual basis, autochthonous autotrophy provides, through an algal–grazer food web pathway, the trophic basis for most metazoan productivity for the riverscape as a whole, but allochthonous organic matter may be more important for some species and seasons and in shallow, heavily canopied headwaters; however, a collateral and weakly linked, decomposer food pathway (the microbial–viral loop) is primarily responsible (with algal respiration in some cases) for a river's heterotrophic state ($P/R < 1$).

There is a growing body of evidence indicating that autochthonous production is a major component of metazoan secondary production in streams and rivers (Delong and Thorp, 2006). Exceptions will exist as a function of the constraints on autochthonous production within certain FPZs, specifically the light environment, nutrient availability, and patches with physical attributes for supporting higher rates of primary production. Given the importance of locally derived instream primary production, there is the potential that autochthonous production will remain the primary basal resource in disturbed FPZs. System energetics could still change despite a continued reliance on autochthonous production as a result of changes in attributes of an FPZ. These changes could include shifts in the quantity, type (e.g., benthic vs pelagic microalgae), nutritive value (e.g., shift from green algae to cyanobacteria, affecting availability of omega-3 fatty acids), and timing of autotrophic production. These may appear to be of little consequence, but examination of each change will demonstrate their implications on overall system productivity and ecosystem health. If the quantity and quality of autotrophic production in a disturbed FPZ is inadequate to sustain resident metazoan consumers, then they must rely more on generally lower-quality allochthonous production or their diversity and density will diminish.

All forms of autochthonous production are not equal in their energetic and chemical value for metazoan productivity. Different species of algae possess different combinations of fatty acids, which define much of their energetic value. Diatoms and chlorophytes possess large amounts of polyunsaturated fatty acids (PUFAs), which are important building blocks of organic molecules, which many consumers cannot synthesize (Brett and Müller-Navarra, 1997). Conversely, cyanobacteria possess very little of these essential fatty acids making them a lower-quality food source for consumer growth. Other quality changes, linked to the ability of primary consumers to use certain types of autotrophs, can also change system energetics. Evidence suggests that microalgae, both benthic and planktonic, are a major basal food resource in rivers (Hamilton *et al.*, 1992; Delong and Thorp, 2006), whereas filamentous algae do not appear to be used extensively (e.g., Delong *et al.*, 2001). Therefore, changes in the size distribution of autotrophs as a result of a reconfiguration of an FPZ would affect trophic condition. Nutrient enrichment of arctic streams promoted the growth of filamentous algae and bryophytes (Harvey *et al.*, 1998). These macroautotrophs could potentially limit microalgal production, thereby shifting resource use by primary consumers and changing metazoan production. As an important determinant of nutrient dynamics, changes in hydrological retention, whether in high-residence time patches in the riverscape or the periodically connected aquatic and terrestrial habitats of the floodscape, have implications for the quality and abundance of autotrophic production. Alterations in riverine landscapes that shorten the duration of hydrological retention in the riverscape, such as channelization, will not allow sufficient time for growth and reproduction of longer-lived autotrophs important to the food web. Conversely, expansion of the duration of hydrological retention has the potential to lead to nutrient depletion and to promote increases in the abundance of low-quality resources like cyanobacteria.

The timing of resource availability can be altered by disrupting historic hydrologic patterns through modifications in the riverscape or watershed. Uehlinger (2000) examined two regions of the River Thur, a prealpine river in Switzerland where the study areas appear to represent two distinct FPZs. Although the two contrasting sites described below directly pertain to system metabolism and only indirectly to food webs, we suggest that the implications for resource availability are clearly evident. Hydrologic, bed-moving events at a study site laterally constrained by bedrock and which featured a high slope (0.60%) shifted the system more toward heterotrophy due to a sizeable reduction in gross primary production (GPP). In contrast, smaller decreases in GPP at a site with a large hyporheic zone and lower slope allowed this area to remain autotrophic. Increases or decreases in the magnitude or periodicity of spates by anthropogenic activities may elicit similar responses. Modifications that increase the rate of runoff, such as urbanization and drainage systems for rapidly removing water from agricultural fields, greatly increase the frequency of flow pulses. Such events remove benthic algae through scouring. Greater frequency of flow pulses also creates shorter intervals of high hydrological retention, thereby reducing the potential for phytoplankton production. Shifts in hydrology, including timing of flood pulses and low-flow intervals, have the potential to divert peak production of autochthonous organic matter to times of the year that do not coincide with periods of maximum-potential secondary productivity. Reduction in the availability of high-quality food sources forces Metazoa to rely on lower-quality, and potentially less available, resources. Regulation and water extractions in the River Murray, Australia, have reversed the seasons when flood pulses and low-flow periods were most common (Thoms and Walker, 1993). This change in the periodicity of hydrological processes is likely to cause a decline in the amount of primary production on an annual basis if optimal flow conditions do not match with environmental conditions for maximum rates of productivity.

Even if total primary production remains constant, community composition and food web pathways would be disrupted by a shift from the preponderance of benthic autotrophy to dominance by pelagic (sestonic) sources or vice versa because of differences in adaptive mechanisms of resident herbivores. Such changes would occur if critical attributes of a species' habitat template are altered markedly. Bank stabilization projects, for example, are intended to mitigate channel-forming processes. Diminished erosion and the presence of bank stabilizing riprap divert in-channel flow and reduce the system's capacity to form vital slackwater zones by either scouring of the channel's margin or deposition of eroded sediments and woody debris further downstream. Riprap provides substrate for increased periphyton growth, increasing rates of benthic primary production in the modified FPZ. Impoundments appear to cause shifts in autochthonous resource consumption in both the impounded waters and the river below a dam from local resources to autochthonous production derived from the reservoir (e.g., Angradi, 1994). These shifts in resources, coupled with other changes in the nature of the FPZ above and below a dam, are often the causal mechanisms responsible for differences in biocomplexity commonly reported in rivers as a result of dam construction. These resource shifts represent a fundamental change in energy flow within the ecosystem.

Channelization reduces hydrological retention in the riverscape through the combined effect of decreased water travel time and the elimination of shoreline sinuosity, thereby effectively creating a constricted FPZ. In this situation, both benthic and planktonic primary production should decline. Loss of hydrological retention impedes the ability of phytoplankton to grow and reproduce, whereas loss of shoreline surface area reduces the amount of surface area available for benthic algal production. Levees also create changes in an FPZ that should impact both planktonic and benthic primary production. In both of these situations, a potential exists for a shift from a food web sustained by autochthonous production to one supported by allochthonous production, but this change will be dependent on the nature of the FPZ present prior to modification and the extent of anthropogenic changes to the system. Delong *et al.* (2001) examined trophic linkages of the Upper Mississippi, lower Ohio, and lower Missouri rivers. Both the lower Ohio and Upper Mississippi rivers have been modified for navigation through the construction of low-head dams (Delong, 2005; White *et al.*, 2005). Although "free flowing," the lower Missouri River has been the most extensively modified of the three. Formerly a wide, extensively braided river, the Missouri has been channelized with its lateral movement and connectivity restricted by levees. Hydrological variability has also been reduced by the construction of several large reservoirs much further upstream (Galat *et al.*, 1998). The Upper Mississippi and lower Ohio rivers have food webs that are supported primarily by living phytoplankton, with some contribution of detrital autochthonous matter. The primary food resource identified for the lower Missouri was C_3 leaf litter from the floodplain forest. Although the historic lower Missouri carried a suspended sediment load even greater than it does today, the original shallow braids and greater lateral and longitudinal complexity of the system prior to modification would have fostered autochthonous production even in the channel margins comparable to that observed in other turbid systems (e.g., some Australian streams: Bunn and Davies, 1999; Bunn *et al.*, 2003). The importance of autochtony is also supported by the historical presence and abundance of many plantivorous fish in the mainstem Missouri, ranging from small cyprinids (*Notropis* spp.) to large paddlefish (*Polyodon spathula*) and bigmouth buffalo (*Ictiobus cyprinellus*) (Pflieger, 1975).

Despite this emphasis on changes in autochthonous production, it is also true that the primary resource for some sections of some riverine ecosystems is allochthonous organic matter. This is particularly true for FPZs that are heavily canopied, creating a light environment that limits autotrophic production. Removal or reduction of the riparian canopy greatly increases the amount of light reaching the stream in addition to reducing input of leaf litter and terrestrial prey, such as insects trapped at the water surface. Both allochthonous and autochthonous production were important basal resources for primary consumers in headwater streams of the Upper Chattahoochee basin (England and Rosemond, 2004). Loss of canopy cover can lead to a shift in resource balance through a reduction in dependence on allochthonous CPOM. Whatever the mechanism, similar responses should be expected when canopy cover is lost or the type of vegetation changes in streams is heavily dependent on allochthonous inputs. Species substitution in the canopy can alter uptake into the aquatic food web because food quality and decomposition rates of CPOM depend on vegetation type (Webster and Benfield, 1986). Although scientists have typically focused on riparian loss from small streams, this can also occur in rivers when the natural banks are replaced by treeless levees and from plowing farmland too close to the channel. This has the potential to directly impact a significant portion of the riverine fish and invertebrate fauna that tend to cluster in shallow-water habitats nearshore for protection and food (Thorp, 1992).

Alteration of an FPZ also has implications on the microbial food web, thereby influencing the nature of heterotrophic processes. As predicted in the riverine productivity model (Thorp and Delong, 2002), studies have indicated a correlation between phytoplankton abundance and bacterial production. Karrasch *et al.* (2001) found a strong positive association between phytoplankton abundance and bacterial production, which, in turn, were trophically linked to heterotrophic flagellates. Bukaveckas *et al.* (2002) noted the same association in Kentucky Lake, an impoundment of the Tennessee River. Given this relationship, changes in autochthonous production by any of the modifications of FPZs as already described for this tenet would have implications for the microbial–viral component of the food web in FPZs where autochthonous production is important for overall system productivity. Other losses of organic matter input, including canopy removal and a reduction in the capacity to store organic matter, would alter the nature of bacterial dynamics, given the importance of the delivery and retention of organic matter in sediments as a substrate for bacterial production (Starzecka and Benarz, 2001; Logue *et al.*, 2004).

Model Tenet 12: Floodscape Food Web Pathways

> Algal production is the primary source of organic energy fueling aquatic metazoan food webs in floodscapes of most riverine landscapes, especially in rivers with seasonal, warm-weather floods; the relative importance of autochthonous autotrophy in floodscapes decreases with a rise in the amount and temporal extent of inorganic turbid conditions.

Although not addressed in discussions of tenet 11 above, it is evident from Chapter 6 that processes in the floodscape are important to energy flow and productivity in both the riverscape and the floodscape (Lewis *et al.*, 2000, 2001; Benke, 2001). Although differences will exist between the riverscape and the floodscape at all ecological levels, it is difficult to discuss the two separately, especially with regard to issues concerning energy flow. An issue in this model tenet is how changes occurring in the riverscape (e.g., dams and levees) alter floodscape food webs. In particular, we focus here on responses of food webs within floodscape water bodies (wetlands, ponds and lakes, and isolated oxbows, billabongs, and anabranches) rather than the impact of the normally terrestrial, floodplain component of floodscapes on biotic constituents.

A critical aspect of any discussion for altered riverine landscapes is the consequences of the loss of energetic linkages between riverscape and floodscape. Construction of levees is certainly the best-known example of an anthropogenic change that eliminates riverscape–floodscape corridors. Processes that diminish the magnitude, timing, and duration of the flood pulse can also degrade or eliminate these linkages without the construction of a physical barrier. Flood-control dams function to reduce flood risk to floodplain developments, with the retention of water by large reservoirs allowing for controlled release of water during prolonged flow pulses rather than flood pulses. The result of this is a reduction or elimination of chemical and biological exchanges between the riverscape and the floodscape. Fragmentation of the floodscape is another consideration in changing the nature of food webs. More subtle changes in the floodscape through deforestation, water storage development, and urbanization, even in the absence of large-scale engineering projects, will modify the hydrological character of the floodscape, changing connectivity within the floodscape, potentially altering organic matter movement and autochthonous production (Thoms, 2003).

Severing riverscape–floodscape linkages has serious implications to trophic dynamics of the riverine landscape. Inundation of the terrestrial floodscape introduces nutrients that allow for enhanced phytoplankton and benthic algal production in shallow, relatively clear waters of the inundated floodplain (Lewis *et al.*, 2000). This autochthonous production is a major component of the food web in temperate and tropical floodplain ecosystems (Araujo-Lima *et al.*, 1986; Winemiller, 2004; Lindholm and Hessen, 2007; Sommer *et al.*, 2007). Primary production on the terrestrial floodscape during inundation may, in fact, be the primary source of energy for growth and reproduction over the course of the year, with other resources, including terrestrial organic matter, providing only maintenance energy for floodplain-dependent taxa after the river has returned to below-bankfull levels (Araujo-Lima *et al.*, 1986). Loss of riverscape–floodscape linkages, therefore, would exclude essential resources to riverscape and floodscape trophic pathways, resulting in a concomitant loss in total system productivity. This may be especially detrimental in riverine ecosystems where lotic productivity is low yet secondary production is unexpectedly high. For example, larval fish in the Okavango inland delta, Botswana, benefit from resources made available as the nutrient-poor river water flows over the nutrient-rich savannah, with the resulting increased aquatic primary production driving zooplankton production (Høberg *et al.*, 2002).

The power of river waters as they flow over the banks and onto the floodscape defines the physical, chemical, and biological character of the aquatic patches of the floodscape. Susceptibility to flooding, based on a combination of distance from the river, elevation, and topography between river and lake, dictates the size, depth, and age of floodplain lakes (Miranda, 2005). Moreover, exchanges with the river influence water quality, species composition, nutrient concentrations, and composition of organic matter (Hamilton and Lewis, 1987; Sparks, 1995; Tockner *et al.*, 2007). With the significance of this relationship on the type of species present and basal resource availability, it should be expected that the elimination of riverscape–floodscape linkages would alter trophic dynamics in floodplain lakes. The major feature seen in floodplain lakes in unmodified riverine landscapes is an increase in phytoplankton and benthic algal production. This response has been documented for dryland rivers (Jenkins and Boulton, 2003) and inundated terrestrial floodplains (Hamilton *et al.*, 1992). Elimination or diminution of the flood pulse, therefore, should be expected to reduce the availability of autochthonous resources, potentially to the point where consumers must shift to alternative resources or face a decline in secondary production and likely reduction in species diversity. Periodic flooding of aquatic habitats of the floodscape initiates a reset of conditions through the introduction of nutrients, organic matter, and biota (Gruberts *et al.*, 2007; Thomaz *et al.*, 2007). Conditions within these aquatic patches are characterized by a predictable temporal pattern influenced by the physical and biological characteristics of each habitat following recession of flood waters. Removal of inundation as a driver of the system, therefore, would shift the control of trophic dynamics to internal factors. Nutrient depletion in the absence of inundation or nutrient replenishment from connection to the river and surrounding terrestrial floodplains are examples of primary mechanisms that can alter trophic dynamics. Nutrient depletion may be indicated by increased abundance of nitrogen-fixing cyanobacteria, which represent a low-quality food resource for primary consumers and translate into less energy available to support higher consumers.

Model Tenet 13: Nutrient Spiraling

Average nutrient spiraling length varies significantly among functional process zones, with spiral length: (i) increasing in a linear fashion directly with current velocity but changing abruptly at thresholds where boundary sediments are entrained; and (ii) decreasing with greater hydrologic retention and/or floodscape storage resulting from enhanced vertical and lateral geomorphic complexity.

As implied by this tenet, hydrological processes, biotic and abiotic retention, and geomorphic complexity are key attributes of FPZs in defining the length of nutrient spirals. Alterations within and among FPZs in any of these attributes, therefore, will elicit a change in spiral length and nutrient pathways. Not coincidentally, the interactions of these attributes and subsequent alterations can work together to alter nutrient dynamics in riverine landscapes. Flow and flood pulses, for example, dictate the rate of movement of nutrients in the water column in addition to nutrient delivery and removal from slackwater zones and the floodscape. These pulses are also responsible for the formation, destruction, and relocation of retentive features in rivers, including determining the distribution of the biotic components of spiraling. The reasoning becomes circular when you consider that man-made changes in geomorphic features change the nature of hydrological processes, with some modifications pushing discharge beyond threshold velocities for nutrient transport. Consider modifications leading to a greater frequency of high discharge events through changes in runoff (urbanization, tile draining of agricultural lands, deforestation, etc.). Functional process zones that have been subjected to these changes will exceed threshold velocities more frequently, leading to greater annual nutrient transport. Spates have the additional effect of removing organic debris dams, aquatic vegetation, organic matter-rich gravel/sand bars, and slackwater sediments, all of which are important sites of denitrification (Schaller *et al.*, 2004; Groffmann *et al.*, 2004). Loss of these retentive features reduces nitrogen uptake, resulting in elongated nutrient spirals with faster downstream transport in impacted FPZs.

Gücker and Böechat (2004) compared ammonium retention for 12 savanna headwater streams of different levels of geomorphic complexity in the Rio Cipó Basin, Brazil. They concluded that nutrient uptake potential and storage were highest in streams with complex morphologies, particularly when highly retentive swampy areas were present in the riverine landscape. Uptake potential in runs, with their low sinuosity and lack of lateral retention, represented the other extreme of the spectrum. This latter example is most representative of FPZs where a geomorphically complex channel has been simplified. Levees and channelization reduce overbank flooding by speeding the rate of water movement through the riverine landscape, but the geomorphic costs are the reduction of channel migration leading to diminution in shoreline sinuosity and reduction in the capacity to form other retentive geomorphic features such as bars, channel braids, and marginal slackwaters (e.g., Poff *et al.*, 1997). The expected response for these and other modifications that reduce in-channel and lateral complexity would be a lengthening of the nutrient spiral.

A severe reduction in nutrient processing rates occurs as a consequence of channelization, levee construction, and other anthropogenic alterations that stabilize flow regimes or reduce the magnitude of flood pulses. This disruption results because the laterally expanded riverscape and various aquatic and terrestrial components of the floodscape can no longer function in their natural role as long-term nutrient sinks via biotic uptake and chemical transformation of inorganic nitrogen and other macro- and micronutrients (Koschorreck and Darwich, 2003; Richardson *et al.*, 2004). Floodplain lakes also act as sinks for riverine nutrients. Riverine nutrient concentrations decreased downstream of floodplain lakes during floods, whereas concentrations in floodplain lakes increased along a 400-km section of the Middle Parana River (Maine *et al.*, 2004). These nutrient deliveries are, in turn, important to the functioning of floodscape lakes by providing a recharge of nutrients for autotrophic and microbial production and liberating nutrients through denitrification and methanogenesis (Carvalho *et al.*, 2003; Nielsen *et al.*, 2004; Rejas *et al.*, 2005).

Elimination of riverscape–floodscape connectivity and loss of geomorphic complexity will reduce nutrient retention, leading to increased nutrient transport and lengthening of the spiral. The presence of a hypoxia zone in the Gulf of Mexico in proximity to the deltaic fan of the Mississippi River is evidence of the highly detrimental impact of altering large stretches of FPZs in this river. Increased nitrogen load to the river from fertilizer runoff from agricultural lands has been identified as the primary agent responsible for the formation of the Gulf hypoxia zone (Rabalais *et al.*, 2002). Modifications of the lower Mississippi River have greatly diminished geomorphic complexity by converting the river into what is in some people's mind hardly more than a deep drainage ditch. Where the geomorphic complexity of the FPZs of the lower

Mississippi River is intact, it is conceivable that there would be a reduction in nitrogen loading to the Gulf of Mexico. Even removal of all but the most vital levees, however, might not be enough to completely eliminate the Gulf hypoxia zone because current estimates of the rate of nitrogen runoff from agricultural lands are thought to greatly exceed the capacity for nitrogen retention through microbial processing (Kemp and Dodds, 2002; Schaller *et al.*, 2004). In essence, we could achieve some reduction in overall spiral length (a worthy goal) but probably not enough to entirely restore the health of the Gulf.

Disturbances to FPZ characteristics described here also lead to changes in the diversity of terrestrial and aquatic plants, microautotrophs, invertebrates, and fish all of which represent storage components in nutrient spiraling. Riparian vegetation provides a buffer at the land–water interface that reduces inflow of nutrients to streams. Reduction in the width of riparian zones or changes in species composition lead to increased nutrient loading (Mander *et al.*, 2000; Lin *et al.*, 2002; Kiffney *et al.*, 2003). Increased autotrophic production is the typical response to the combined effect of increased nutrient loading and greater light penetration through canopy loss, but the threshold for biotic uptake will ultimately be overwhelmed and spiral length will increase. Floodplain vegetation also serves as an important sink and benefit from the availability of nutrients that result from inundation of the floodscape (e.g., Tibbetts and Molles, 2005). Natural riparian vegetation removed higher concentrations of nitrate from groundwater in the floodplain of the Garonne River, France, than did poplar plantations and wet meadows (Pinay *et al.*, 1998). Valett *et al.* (2002) determined that phosphorus concentrations in North Carolina, USA, in the Appalachian Mountains were lower in streams flowing through old-growth hard-wood forests when compared to streams draining second-growth forests. They also noted that standing stocks of debris dams were several orders of magnitude greater in old-growth forests. It is important, therefore, that both species composition and successional status of the floodscape be accounted for in evaluating nutrient spiraling in modified FPZs.

As discussed when describing species changes in impacted FPZs, composition and abundance of Metazoa in altered FPZs will be different from taxa expected in the unmodified FPZ. Changes in metazoan community composition would be expected to impact biotic retention and alter the cycling of nutrients through trophic transfers. Replacement of long-lived species by species with more rapid turnover times, such as midges (Chironomidae), worms (Oligochaeta), or minnows and shiners (Cyprininidae), would create a more rapid return of nutrients to the water column. Nonnative species would also elicit changes in nutrient spiraling, even within a relatively natural FPZ through species replacement and changes in food web-related energy flow. The gastropod *Potamopyrgus antipodarum* is not native to Polecat Creek, Wyoming, USA, where it consumes 75% of the GPP and accounts for 67% of the ammonium demand (Hall *et al.*, 2003). The zebra mussel, *Dreissena polymorpha*, a nonnative species in western Europe and North America, consumes large quantities of phytoplankton (Caraco *et al.*, 1997), locking these nutrients into their tissues for prolonged periods. Zebra mussels also influence nutrient cycling through the production of pseudofeces and feces, which provide a carbon-rich source to promote denitrification (Bruesewitz *et al.*, 2006). Introduction of other long-lived, nonnative planktivores could seriously alter nutrient spiraling, potentially creating limitations in nutrient availability even in instances where nutrient limitations would not be expected.

In summary, the general pattern in modified FPZs is that the nutrient spiral is lengthened as most alterations will reduce retentive mechanisms and/or increase nutrient concentrations beyond the capacity of retentive features, both biotic and abiotic, to absorb and store nutrients. Exceptions will exist, as described for some nonnative species introductions. Functional process zones subjected to water extraction may be another exception, given that water removal will increase hydrological retention. This may, however, be short term with nutrient transport increasing once autotrophic production reaches a maximum point and biotic nutrient uptake peaks.

Model Tenet 14: Dynamic Hydrology

> Naturally dynamic hydrological patterns are necessary to maintain the evolved biocomplexity in riverine ecosystems.

Poff *et al.* (1997) identified five critical components of flow regime that regulate ecological processes in riverine ecosystems: magnitude of discharge; frequency of occurrence of flows of a given magnitude; duration of specific hydrological conditions; timing or predictability of flows of a defined magnitude; and rate of change from magnitude of discharge to another. These components can operate within different temporal scales defined by Thoms and Sheldon (2000): flow regime, the generalization of flow behavior, which reflects a scale of hundreds of years; flow history, the sequence of floods or droughts as measured over 1–100 year; and flood pulse, flood events occurring within a time span of <1 year. These attributes reflect the dynamic nature of hydrological processes in riverine landscapes and, ultimately, define the realized biodiversity and ecosystem function of an FPZ (e.g., Thoms and Parsons, 2003). Substantial changes in any of these five hydrological components, particularly across the spatial scales relevant to flow regime and flow history, will alter attributes of an FPZ, thereby shifting biocomplexity to a different state.

Poff *et al.* (1997) provided excellent examples of the hydrological and geomorphic responses to different types of anthropogenic activities that modify flow regime in addition to ecological responses to alterations of the five components of natural flow regimes, and we have discussed in this chapter the implications of these changes to the physical, chemical, and biological attributes of FPZs. We propose that changes in natural dynamic hydrological patterns will reduce biocomplexity by: (i) creating a new FPZ that is fundamentally different from the original FPZ; (ii) homogenizing adjacent FPZs into a single FPZ; or (iii) reducing the role of lateral and longitudinal complexity in FPZs where connectivity to lateral and longitudinal components is critical to biocomplexity. The implications of anthropogenic modifications on the dynamics of system complexity and connectivity are described in more detail below in tenet 16.

The lower Illinois River, USA, exhibited a predictable and protracted spring flood pulse prior to water diversions and the construction of navigation dams. This river flows through a valley once occupied by the Mississippi River, placing a river of much lower discharge into a valley 8–11 km wide (Delong, 2005). Prior to its modification, the wide floodplain of the Illinois River, USA, was inundated by flood pulses lasting up to 6 months as a result of the combined effect of a river gradient of only 2 cm/km and the *hydraulic dam* effects of the Mississippi River slowing flow far upstream of their confluence. A predictable summer low-flow period and only sporadic flow pulses characterized the hydrology of the unaltered Illinois River. The story, however, is very different in the modern Illinois River (post-1900). The hydrology of the lower Illinois River today is much less predictable, with flood pulses occurring more frequently and at higher magnitudes than premodification conditions (Sparks *et al.*, 1998). Protracted low-flow periods are of shorter duration, while flow pulses occur with greater frequency. Further change is evident in lateral exchanges as well. The Illinois River has been subjected to extensive levee construction to protect agriculture. The river is now considered an aggrading system with a high suspended sediment load because its levees limit the amount of floodplain surface available for sediment storage. Instead, sediments are deposited between the river channel and the levee, creating a floodplain terrace on the riverside of levees that is often 4 m higher than the disconnected floodplain on the protected side of the levee (Delong, personal observation). The result is that the physical and hydrological nature of the lower Illinois River is considerably different, taking on the form of an FPZ markedly different from its earlier form.

Changes in the hydrological character of the Condamine-Balonne riverine ecosystem, Queensland-New South Wales, Australia, have been examined using the techniques to define

FPZs. Like many of Australia's rivers, water extraction for irrigation is a major characteristic of the system's modern hydrology. To ascertain how flow dynamics of the system has changed, Thoms and Parsons (2003) used a suite of hydrological variables reflecting flow regime, flow history, and flow pulse from the periods before and since water extractions began. Historical, preirrigation data identified six hydrological zones in the riverine ecosystem. Application of the same process of data covering the period of water extraction revealed that both the midzone and the anabranch zone, the latter occurring toward the mouth of the river, had increased in size, with both effectively absorbing an entire hydrological zone that is no longer evident in the modified Condamine Balonne River. Pegg and Pierce (2002) observed similar homogenization of hydrological zones as a result of the profound changes wrought on the Missouri River, as well as the formation of FPZs represented by the reservoir and interreservoir zones in North and South Dakota, USA. Given that hydrological characteristics are a key component in defining FPZs, both of these systems now possess an arrangement and abundance of FPZs markedly different from what existed in their undisturbed state. The outcome from these changes would be that biocomplexity will be markedly different and reduced in both structure and function in response to the loss of FPZs. Even the addition of new FPZs will not replace this lost biocomplexity, given these new FPZs represent pronounced state changes relative to conditions present during the genesis and development of the riverine ecosystem.

Model Tenet 15: Flood-linked Evolution

The frequency of flood-linked life history characteristics increases directly with the seasonal predictability of floods and their concurrence with periods of maximum system primary productivity.

Floods serve as a major driver in the reproductive strategies, habitat use, and life histories of biota across the riverine landscape. Their influence is seen in the number of strategies that have evolved, including: avoiding periods of flooding, as in the case of synchronized emergence of terrestrial adults (e.g., Lytle, 2002) or through the maintenance of high levels of genetic diversity allowing linkages between episodically isolated subpopulations (e.g., Carini *et al.*, 2006). It should be expected that anthropogenic changes, particularly in the timing, magnitude, and duration of flood cycles, would elicit a genetic response through changes in conditions that influence fitness of life history traits and potential availability of suitable habitat and energy resources.

Fish provide some of the best examples of the relationship between flood and life history strategies. Migratory behavior of many fish is triggered by flood or flow pulses, with the timing of these linked to finding spawning sites, juvenile migration to adult habitats, or escape to refugia (Sparks, 1995; Pringle *et al.*, 2000). Loss or changes in the timing of flow-linked cues alter fish behavior, decrease reproductive success, or increase mortality through greater exposure to predation or loss of access to refugia required during periods when environmental conditions become harsh (e.g., Raibley *et al.*, 1997). Some life history strategies, such as buoyant eggs, are dependent on floods to carry them into the floodscape or in-channel refugia where they can mature and develop into young fish. Similarly, the structures that alter flood dynamics also serve as barriers. Dams prevent upstream and downstream migration, whereas levees eliminate the corridor between riverscape and floodscape created by floods (Pringle *et al.*, 2000; Barko *et al.*, 2006). Compounded by this is the critical role of floods in the formation of the mosaic in physical habitat evident in flood-driven riverine landscapes. Loss of the capacity to shape and reconfigure the landscape through the actions of floods further diminishes the relevance of flood-linked traits and reduces success of species that evolved under the influence of flood processes.

Flood-linked evolution is not limited to fish. Plant species in the riverine landscape have evolved similarly with some species adapted to survive flood pulses and others evolving life history strategies to avoid periods of flooding. Reduction or elimination of floods opens the floodscape to colonization by native species with little or no tolerance to flooding and by nonnative species. Flow stabilization has greatly reduced the frequency of flooding in the floodplain of the Rio Grande, New Mexico, USA, affecting various tree species. For example, cottonwoods, *Populus deltoides*, have declined precipitously to be replaced in dominance by salt cedar, *Tamarix ramosissima*, and Russian olive, *Elaeagnus angustifolia* (Molles *et al.*, 1998). Aquatic invertebrates have adapted a number of reproductive and habitat-use-related strategies as a consequence of inhabiting flood-influenced systems. Aquatic insects in the flash-flood-prone streams of montane regions of the Chihuahuan Desert, Arizona, USA, have evolved a strategy of emerging before the long-term mean arrival of the first seasonal flood (Lytle, 2002). This strategy becomes moot in the absence of the seasonal flood, opening up the system for potential colonization by aquatic insects that are not adapted to these conditions, with the potential loss of adapted species through increased biotic interactions.

Patterns of flooding can also influence the overall genetic diversity of the fauna found within a given FPZ. Dryland rivers in western Queensland, Australia, form isolated water holes during periods between floods, with fauna crammed into the few remaining pools for extended periods. Genetic studies of the shrimp *Macrobrachium australiense* and the gastropod *Notopala sublineata* have determined that, despite extended periods of isolation, a high degree of genetic similarity exists and that genetic exchanges evident between water holes maintain the genetic diversity of these two species (Carini *et al.*, 2006). Loss of the episodic floods that link water holes would lead to genetic isolation, greatly diminishing the genetic diversity of these two invertebrates and, as a result, their potential to adapt to later natural or anthropogenically driven changes in environmental conditions.

Model Tenet 16: Connectivity

> Biocomplexity generally peaks at intermediate levels of connectivity between the main channel and lateral aquatic habitats of the riverine landscape, but the relationship varies substantially among types of connectivity, evolutionary adaptation of taxa to flowing water, and functional processes examined.

A recurring theme in Chapters 6 and 7 has been the overall importance of connectivity throughout a riverine ecosystem. We have defined connectivity spatially as a mechanism where linkages are established between: (i) hydrogeomorphic patches within the riverscape; (ii) the riverscape and the floodscape; and (iii) patches within the floodscape. Connectivity means a linkage between patches, with each patch representing a potential key point at the level of individual, population, community, or ecosystem. There is likely, however, to be a minimum and maximum threshold for spatial connectivity within an FPZ, whereby biocomplexity would be maximized somewhere between these thresholds. The question then for spatial connectivity and anthropogenic changes is in which direction does the disturbance push ecological processes along this gradient?

Another consideration is the implications in changing the timing or duration of connectivity within a given FPZ. Studies on the Upper Mississippi River have determined that removal of nitrogen from the river by denitrification is determined, to a large degree, by the timing of the delivery of water with high nitrate concentrations to patches containing high concentrations of organic carbon in sediments that are prone to anoxic conditions (Strauss *et al.*, 2006). Dryland rivers, such as the Darling in southeastern Australia, are characterized by a *boom and bust* ecology. Booms in primary and secondary productivity as well as colonization by invertebrates and vertebrates are dictated by the magnitude and timing of reconnections both within the riverscape and between riverscape and floodscape (e.g., Jenkins and Boulton, 2003). Changing the time at which connectivity occurs to periods where environmental conditions are not conducive for increased biological activity or does not coincide with migratory and dispersal periods of organisms dependent on this productivity would seriously reduce biocomplexity within an FPZ where connectivity is an important facet of system processes.

Although not discussed explicitly in this chapter, disturbances that change the type and arrangement of hydrogeomorphic patches can alter the nature of connectivity within either the riverscape or the floodscape. This is applicable at any spatial scale, but may be best addressed by looking at the level of functional unit (Thoms and Parsons, 2003). Loss of diverse habitats, such as backwaters, secondary channels, and marginal slackwaters, leaves only habitats strongly connected to the main thalweg in the riverscape. A similar effect would be seen within the floodscape through the conversion of land to agriculture and the draining of wetlands and lakes. In these situations, spatial connectivity is actually increased since the delineations between hydrogeomorphic patches are essentially lost. Examples are given throughout this chapter as to how loss of hydrogeomorphic patches leads to the decreases in population densities and lower species diversity through replacement of habitat specialists with macrohabitat generalists. Ecosystem function would be simplified, in essence, because the range of resources available to consumers at all trophic levels would be limited and greater connectivity would likely reduce the abiotic and biotic retention of nutrients. The prediction with greater spatial connectivity within the riverscape or the floodscape is that a reduction in critical components would create a system with lower biocomplexity.

Connectivity within the riverscape is also shaped by hydrological conditions, particularly by the influence of hydrological variability. Natural hydrological variation exerts considerable influence on the degree of interaction between hydrogeomorphic patches within the riverscape; thus many of the ecological responses are similar between the two. A mechanism that is likely to be important to biocomplexity is the relative importance of abiotic vs biotic controls of distinct hydrogeomorphic patches within the riverscape. Baranyi *et al.* (2002) proposed that the length of intervals between flow pulses would dictate processes controlling plankton communities. Abiotic controls associated with the main flow would define biocomplexity during periods of high connectivity. A longer interpulse period diminishes the thalweg influence and potentially provides a greater opportunity for biotic controls to influence ecological processes. The magnitude of hydrological variation will, obviously, vary as a function of the type of FPZ encountered. The amount of change in biocomplexity, therefore, will be dependent on the degree to which hydrological variability deviates from expected patterns for an FPZ.

Connectivity between riverscape and floodscape has been more thoroughly examined, and it is widely considered to be a major driver of ecological processes in floodplain rivers (Junk *et al.*, 1989; Poff *et al.*, 1997; Lewis *et al.*, 2001), including biodiversity (Galat *et al.*, 2005; Tockner *et al.*, 2000; Robinson *et al.*, 2002). Numerous examples of the impact of this interaction at all levels of the ecological hierarchy are evident throughout Chapters 6 and 7. What must be remembered, however, is that the timing, magnitude, duration, frequency, and rate of change during connectivity will dictate the nature of biocomplexity under natural flow regimes (*sensu* Poff *et al.*, 1997). The degree to which any of these parameters are altered, therefore, will define the changes seen within a modified FPZ. The examples related to in this chapter have dealt primarily with the implications of reducing or eliminating riverscape–floodscape connectivity. It is possible, however, to have too much connectivity between the two, with definite implications on ecosystem processes. The annual hydrological pattern of the Illinois River, once characterized by a prolonged and predictable flood pulse, fluctuates widely with multiple flood pulses throughout the year (Delong, 2005). Frequent inundation of the floodplain has prevented plant growth for species, such as the millets *Echinochloa* spp. and decurrent false aster *Boltonia currens*, that are adapted for the long interpulse period (Ahn *et al.*, 2004; Smith *et al.*, 2005). Prolonged connectivity should be expected to illicit similar detrimental effects on other ecological processes and biocomplexity on the floodscape.

Model Tenet 17: Landscape Patterns of Functional Process Zones

> Biocomplexity in riverine ecosystems is significantly influenced by the diversity, spatial arrangement, and character of the functional process zones.

This chapter has focused on processes occurring within an FPZ and how anthropogenic disturbances affect them; however, what happens if the distribution, diversity, and arrangement of FPZs are altered by major changes in attributes of an FPZ(s), insertion of a new FPZ(s), or the elimination of an FPZ(s) at the level of a riverine ecosystem? These questions have rarely been addressed and are not likely to be tackled in the near future because of major logistical and budgetary challenges. Therefore, the following discussion is very tentative and highly speculative at this point. We explore questions on how the type and the arrangement of patches affect ecological processes at two spatial scales: within an FPZ and at the valley/basin scale of an FPZ.

A major supposition of this tenet is that diversity and spatial arrangement of patches within an FPZ will be a principal determinant of biocomplexity within that FPZ. Moreover, ecotones within an FPZ and those between different FPZs will define the level of biocomplexity within an FPZ and, ultimately, within a riverine ecosystem. What this creates, then, is a scale of differing levels of biocomplexity in undisturbed FPZs as a function of the spatial complexity of the FPZ. Despite possessing different levels of biocomplexity, it should be expected in any FPZ that changing the spatial arrangement and diversity of patches within any FPZ will impact biocomplexity through reduction in the capacity for transfers of biota, nutrients, and energy between patches as well as the loss of ecotones that may prove to be of critical importance in sustaining biocomplexity (see tenet 3). We have previously proposed in this chapter that redistribution or elimination of patches within an FPZ will reduce structural and dynamic aspects of biocomplexity if these changes are pronounced.

The RES further proposes that the types of FPZs present in a riverine ecosystem will establish the structural attributes and functional processes of biocomplexity for the entire riverine landscape. What happens, then, when FPZs are modified or changed to the point of representing a different type of FPZ or a new FPZ is introduced? What are the implications to the biocomplexity of the Condamine Balonne riverine ecosystem where water extraction has caused the expansion of one FPZ and the merging of two FPZs into a single new FPZ leaving five FPZs where there were once six (Thoms and Parsons, 2003)? The same question could be asked for the mainstem Missouri River where impoundment for flood control and channelization have reduced what was once a spatially and temporally complex system to a limited number of regions where hydrological characteristics are now similar for hundreds of kilometers (Pegg and Pierce, 2002). Based on the prediction of this tenet, biocomplexity of modified riverine ecosystems will reflect the new spatial arrangement, diversity, and character of the FPZs. Some changes, such as an impoundment, may serve to increase biocomplexity, at least locally, through the introduction of new set of biota, physicochemical characteristics, and productivity dynamics (Petts, 1984; Casas *et al.*, 2000; Grubbs and Taylor, 2004). Other alterations will reduce network biocomplexity as the internal diversity of FPZs is reduced and entire FPZs are lost (Havel, 2000; Armitage *et al.*, 2003; Slipke *et al.*, 2005; Strauss *et al.*, 2006). Consider the implications to species diversity in a riverine ecosystem. Transitions from one FPZ to another form ecological nodes where species diversity should be maximized through the merging of smaller-scale patches within each patch (see tenet #3). Addition or loss of FPZs will change the attributes of nodes derived by transitions from one FPZ to another, with changes in diversity swinging either in the positive or the negative depending on the level of physical complexity at FPZ–FPZ nodes. Changes of this nature, therefore, would have ramifications for biocomplexity of the riverine ecosystem. Even the construction of an impoundment should reduce species diversity at the level of the riverine ecosystem as the location of the dam may eliminate critical habitat and migratory pathways for species.

A key element of the RES is that FPZs are large hydrogeomorphic patches whose location is dictated by geomorphic and hydrological attributes rather than by where they occur on a clinal gradient. This is not to say, however, that the location of an FPZ does not exert an influence on FPZs downstream or even immediately upstream. It is the arrangement of FPZs that will define processes at the scale of a riverine ecosystem. What does the arrangement of FPZs mean to biocomplexity at the level of an altered riverine ecosystem, especially considering that more than one human impact is likely to have occurred? For example, how important is the location of a channelized region and a high dam? Both change nutrient dynamics differently. Nutrient transport increases in channelized systems through the loss of hydrological and physical retention. Conversely, retention is high in the new FPZ created by the impoundment. Whether the impoundment is located above or below, the channelized FPZ would, therefore, have implications for nutrient spiraling within the entire riverine ecosystem.

8

Practical Applications of the Riverine Ecosystem Synthesis in Management and Conservation Settings

Introduction
Revisiting hierarchy and scales
Application of functional process zones
River assessments and the importance of the functional process zone scale
Determining environmental water allocations
Summary

INTRODUCTION

The use of modern scientific concepts and methods in managing riverine landscapes is a measure of sound science and should always be employed when making decisions affecting the environment or natural resources (Cullen, 1990). To manage riverine ecosystems effectively and efficiently, the basic processes governing ecosystem structure and functioning must be understood at appropriate scales. The scientific and natural resource management literature contains numerous examples of the drastic consequences of inappropriate attempts to conserve and rehabilitate riverine ecosystems without knowledge of their basic structure and function (Boon *et al.*, 2000).

Riverine scientists have contributed to the turbulent field of environmental management – the interface between science and policy – in five main areas related to conflicts in water resource use (Thoms and Sheldon, 2000).

- Descriptions of what exists in ecosystems and identification of key processes and functions.
- Diagnosis of past environmental damage and the present condition of the resource followed by the identification of causes and consequences of the disturbance.
- Prediction of the resource's capability to support various functions and identification of possible hazards, environmental uniqueness or values, and probable ecological effects of specific resource uses.
- Prescription or recommendations on the requirements to maintain the resource within acceptable limits of change. (This is illustrated in this chapter through the development of the ecosystem approach.)
- Advice on formulating management actions and policies.

The relationship between scientists and the people responsible for environmental management and policy is in a state of flux. Identifying issues associated with the impact of humans on river systems was traditionally a primary focus of many river scientists. Within this arena, river managers and policy makers often see scientific knowledge as a commodity that can be purchased as demands arise, while river scientists viewed the fruits of their research endeavors as part of an ongoing search to understand and predict the structure and functioning of riverine ecosystems. However, scientific input in the natural resource management and policy arena has expanded beyond the early notion of writing a scientific publication in the hope that policy makers would read it and then translate its findings into appropriate action (Coates, 1984). River scientists are now involved in many more stages of the overall decision-making process (Knuepfer and Petersen, 2002). A collaborative movement is now emerging in which teams of scientists, managers, and policy makers work collectively to define the problems, generate relevant questions, and provide solutions. The river research program at the Kruger National Park in South Africa is an excellent example of this collaborative process. Previous research and management activities for the park's rivers were principally driven by individual science and management disciplines. Although the program generated abundant data and many scientific and management reports, there was a disjunct between the scales at which most research transpired and the need of the natural resource managers. This limited the application of much of the science to the management of the rivers within the Kruger ecosystem (Biggs, 2003). Fortunately, a radical change in the management and conservation of the park developed as a consequence of the positive effects from a breakdown of disciplinary boundaries and adoption of both a multiscalar ecosystem focus and an objective hierarchy. Research and management moved from a situation where the park was treated as a closed, steady-state ecosystem to one that focused on dynamic landscape heterogeneity. Biodiversity conservation was considered paramount and was seen as dependent on the shifting mosaic of abiotic and biotic factors that create heterogeneity (Rogers, 2003). The "Kruger Experience" (Du Toit *et al.*, 2003) illustrates that an understanding of the variability of biophysical processes operating at different spatial and temporal scales is vital for the development of strategies in rehabilitation and conservation.

REVISITING HIERARCHY AND SCALES

The complexity of river systems challenges many traditional scientific methods. Their multicausal, multiple-scale character limits the usefulness of the conventional reductionist-falsification approach, except when applied at very small scales and within limited domains. Hierarchy theory (see fuller treatment in Chapter 3) provides a framework for interpreting river complexity and for understanding pattern and process at multiple scales (Dollar *et al.*, 2007; Parsons and Thoms, 2007). A river hierarchy is a graded organizational structure, and a particular level (or holon) within this system is a discrete unit of the level above it and an agglomeration of discrete units of the level below it. All levels within a hierarchy exert some constraints on lower levels (O'Neill *et al.*, 1986), especially the one immediately below. Lower levels, conversely, can influence the structure and functioning of higher levels. Hence, the downward constraints and upward influences of a level are primarily encapsulated by the characteristics and properties of the levels immediately above and below. The simultaneous operation of processes at different levels, within particular contextual constraints, gives rise to emergent properties. A level within the hierarchy can be characterized by a scale (O'Neill, 1989). Scale defines physical dimension, especially in terms of grain size and extent, whereas grain size describes the smallest interval in an observation set or the smallest scale of influence of an ecosystem or process driver (Rogers, 2003). Extent is the total area or duration over which observations are made and is the largest scale at which a disturbance or process driver exerts influence on the system. Grain size and extent define the upper and lower limits of resolution of a level within a hierarchy. As a consequence, scale determines the units appropriate for the measurement of variables associated with each level of a hierarchy. For example, two river systems may both represent the same level of organization in a geomorphological hierarchy and can be characterized by the same variables (such as channel density, channel width, and interchannel spacing). If the spatial extent (scale) or scale of focus of the two river systems differ, however, appropriate units for measurement must also differ (Dollar *et al.*, 2007).

Surprisingly enough, very few examples of the explicit use of hierarchy that are relevant to management issues are present in the scientific literature, with a few notable exceptions (Parsons *et al.*, 2003, 2004a; Thoms *et al.*, 2004, 2007; Dollar *et al.*, 2007; Hughes *et al.*, 2008; Parsons and Thoms, 2007). The framework used by Hughes *et al.* (2007) is a useful illustration of how hierarchy theory can link science to management. In this study of the use of large wood for restoring riverine habitat for native fish in the River Murray (Australia), patterns in the nature and distribution of large wood were examined at three hierarchical scales of river morphology: FPZ, functional set, and functional unit. The character and distribution of large wood was uniform at the larger zone scale because the conditions of stream energy were relatively uniform. However, strong associations occurred between lower-level functional sets (straight or bend sections of river) and functional units (12 quadrants within each functional set). The character and distribution of large wood was distinct at these scales because stream energy differed both between straight and bend morphologies and between inner- and outer-channel functional units. Thus, functional sets and functional units are important levels of organization for large wood and the main scale of focus for the restoration of native fish habitat in the River Murray. However, studies by Boys and Thoms (2006) have noted significant differences in the character and distribution of large wood between FPZs, suggesting a more flexible approach to this aspect of river management at the larger catchment scale or river network scale.

The Relevance of Scale in River Management

Selection of the appropriate scale(s) of scientific investigation to underpin management and conservation becomes even more important within an interdisciplinary setting, especially with an increasing trend to manage rivers as ecosystems (Thoms and Parsons, 2002). This requires a holistic, interdisciplinary approach that simultaneously considers the physical, chemical, and biological components of the riverine landscape. Each discipline views river systems from a spatial and temporal perspective, but one of the main impediments to the further expansion of interdisciplinary science and management is the mismatch of scales between disciplines. Interdisciplinary science involves the "... explicit joining of two or more areas of understanding into a single conceptual–empirical structure" (Pickett *et al.*, 1999). Integration of disciplines in this way can be accomplished along additive or extractive lines. The additive case is where two areas of study are combined, more or less intact, into a new composite understanding. In the extractive case, in contrast, different areas of study provide components that are fused to yield a new understanding. Both processes can be used in river science, depending on the nature of the problem at hand and the state of knowledge in the different disciplines. Through the use of hierarchical and integrative frameworks, such as that proposed by Dollar *et al.* (2007), issues of scale can be appropriately reconciled. Such frameworks and the approach proposed in this book allow one to match the description of river form (in the context of a particular management problem) with appropriate riverine ecosystem processes, so that phenomena can be explained at appropriate spatiotemporal scales. This facilitates the understanding and prediction of the response of patterns to processes and the influences of patterns on processes. As pressures increase on the environment, the incentives to manage rivers as ecosystems will continue to build (Palmer and Bernhardt, 2006). This provides a basis for extractive studies in science and science–management exchanges that will potentially bridge the gap between the traditional subject boundaries within science and between science and management (Hannah *et al.*, 2004).

A limited number of environmental management strategies adequately identify the part(s) of the riverine landscape that can or need to be managed, and even fewer have invested in the scientific knowledge required to manage at the appropriate scales (Wissmar and Bisson, 2003). Efforts to gain knowledge on the structure and functioning of riverine landscapes have been thwarted in several ways. First, too few environmental managers acknowledge the hierarchical complexity of rivers and appreciate that different physical and biological patterns within the riverine landscape result from processes at different scales. By incorporating multiple environmental scales in decision-making processes, managers can separate local from systemic influences on river functioning. Second, river conservation, management, and rehabilitation activities are commonly undertaken at the site scale. Thus, the probability of failure increases when planning and design projects are undertaken with inadequate attention to influences of the broader riverine landscape on river reaches or sites (Kondolf and Downs, 1996). Third, there is often a mismatch of scales between needs of the river manager, who is increasingly directing conservation and management efforts at the landscape or catchment scale, and the river scientist, many of whom continue to undertake research at smaller scales. Responsible environmental management requires a good understanding of pattern and process at multiple scales within the riverine landscape.

Selection of appropriate scale(s) for investigating pattern and process in the study and management of riverine landscapes can be accomplished in several ways. In many instances, the scale of investigation or focus is preset, either directly by management actions or by the context of the issue at hand. Alternatively, a large number of river attributes can be analyzed simultaneously in a search for clusters of potentially important spatial or functional variables. Spatial variables can be incorporated into decisions on where to undertake management actions, while functional variables can assist in identifying which processes need addressing by management. A third approach combines elements of both (Parsons *et al.*, 2004b). In situations where scales or levels of organization have not been predetermined for whatever reason, allowing scales to self-emerge may be appropriate. In a study of the distribution of in-channel habitats, for example, Thoms and Maddock (unpublished data) used multivariate statistical analyses to assess the distribution of habitat patches at various scales. In that study of the Cotter River (Australia), no groupings of in-channel habitat were evident at the 50-m scale, but six habitat groups emerged at the 100-m scale. These were dominated by combinations of runs, glide, and pools even though a clear spatial dimension to the distribution of these groups was not evident along the river. A spatial dimension to the habitat organization became apparent only at the 500-m and 1-km scales, with three distinct river zones emerging from the statistical analyses. Each river zone was differentiated by a unique combination of in-channel habitat. The ecological relevance of this approach was demonstrated through its application in defining management actions to eliminate fragmentation in prime habitat for an endangered native fish species, the Macquarie perch (*Macquaria australasica*). Key locations for habitat enhancement were determined along with an assessment of the likelihood of success of reintroducing native species and the effects of environmental flows on fish distributions. A growing body of evidence favors this approach (Thoms *et al.*, 2007; Parsons and Thoms, 2007; Thoms and Maddock, unpublished data), which provides a framework for allowing scales to self-emerge.

Focus on Catchment-Based Approaches to Management

Although early attempts at river management and conservation emphasized site or short reach scales and ignored rivers as ecosystems, catchment scale or landscape approaches are gaining popularity because of increasing emphasis on biodiversity issues. This has been mirrored by an expanding literature base describing the character and processes of broader components of large rivers. A common theme of this trend has been to highlight the highly variable physical, chemical, and biological nature of larger systems and to view riverine landscapes as dynamic and somewhat mercurial ecosystems. The previous and sometimes ongoing reluctance to deal with whole catchments or entire river systems probably stems from the common paradigm that complexity increases with spatial scale (Wiens, 1989; Levin, 1992). Larger-scale ecosystems can be characterized by: (i) higher numbers and types of components; (ii) more interconnections between components; (iii) the presence of both positive and negative feedback loops; and (iv) more interactions between levels of organization compared to those in smaller ecosystems. Although large riverine landscapes are indeed more complex than smaller river reaches or sites within rivers, their analysis and management can be accomplished with proper application of hierarchical scales. Unfortunately, the scientific community has had a love–hate relationship with large or entire river ecosystems. In spite of the curiosity about their unique physical, chemical, and biological patterns and functions (Davies and Walker, 1986), these systems have frustrated some scientists because they do not always conform with contemporary, empirically derived models of river function (Phillips, 1995; Sheldon and Thoms, 2006a). Larger riverine landscapes have been described as being in a nonequilibrium state (Chen *et al.*, 2001; Gupta, 2002) and *intractable*. As a result, they have tended to be less popular to study than either smaller systems or components of river ecosystems. Many of the issues that have thwarted the study of entire riverine landscapes can be overcome with applications of methods and principles at the correct scale. The RES provides a framework for accomplishing this.

APPLICATION OF FUNCTIONAL PROCESS ZONES

A systematic means of identifying assessment targets is essential for those agencies responsible for the conservation, rehabilitation, and management of entire riverine landscapes. The appropriate scale must be included in all actions related to prioritization of areas for environmental action, determination of the feasibility of various options, and establishment of future policies appropriate for biophysical attributes (Stanley and Boulton, 2000). Conservation and management efforts often fail because of the mismatch between the focus of efforts and the efforts themselves. The traditional approach of river conservation and management focusing at the site or small reach scale has given way to whole-system perspectives, especially in the area of river assessment or auditing. In many countries, policies to maintain environmental values rely on information reported at regional and/or national levels. This information is used to answer questions such as the extent and condition of natural resources, how and where the environment is changing, and factors associated with observed changes (Olsen *et al.*, 1999).

Prioritization for Conservation Purposes

Deciding where to invest limited resources in river conservation can be a challenging task (Schofield *et al.*, 2000). It is important to consider questions beforehand such as the following: Which sites are in greatest need of conservation or rehabilitation? How does the response in different sites relate to the relative amounts to be invested? Many river conservation plans have been instituted with little knowledge of the structure and function of the system (Hooke, 1988), and detailed analyses of responses in relationship to priorities are rarely undertaken (Kondolf and Downs, 1996). Determining action targets for sections of a riverine ecosystem must incorporate from the beginning a general understanding of all sections of riverine landscape. Unfortunately, little agreement exists on ways to evaluate sections of a river or even the entire riverine ecosystem for conservation purposes. It is all too common to undertake a river survey to assess river condition and from this determine the appropriate sections of need (e.g., Norris *et al.*, 2007). This approach relies on some form of judgment about what can be achieved.

Many large-scale frameworks for river management rely on judgment of the conservation value of riverine ecosystems (Davies *et al.*, 2004). The use of river characterization has a role in assessing the conservation value of aquatic ecosystems. The approach advocated in the RES notes that riverine ecosystems are composed of repeatable FPZs – large tracts of river with similar physical form and inherent abiotic and biotic processes. The RES not only provides a consistent means of river *typing* at the catchment scale but also outlines a framework for determining a value to sections of a river and even the entire river network. It can be a significant aid in several ways when setting priorities for river management and conservation. For example, FPZs are distinct sections of riverine ecosystems that can serve as spatial filters for setting management and conservation targets. Because of the hydrogeomorphic distinctiveness of each FPZ, they differ in the nature of their riverscape and floodscape and consequently possess unique floral and faunal communities. Protecting an FPZ or sections of them allows the range of habitat used by aquatic biota to be conserved.

The spatial pattern of FPZs allows the abiotic and potentially biotic attributes of a riverine ecosystem to be determined. Knowledge of the distribution of FPZs is especially useful where biotic data are nonexistent or deficient. In this situation, catchment-based conservation or management goals could focus on maintaining a representation of FPZs within a river network. Targets could be set to ensure that adequate lengths of each or certain FPZs within a river are conserved. Indeed thresholds of river lengths could be introduced based on the amount of river habitat particular to each FPZ. In addition, information on the distribution and character of FPZs could be combined with biotic information for each FPZ. A conservation or management priority could then be allocated for each FPZ as determined by their relative abundance, in this case river length. The incorporation of information on the distribution and character of FPZ into algorithms of replicability and complementarity, which often form the basis for systematic conservation planning in terrestrial ecosystems (Margules and Pressey, 2000), would be a major step forward for the conservation of riverine ecosystems. A similar type of approach has been suggested by Turak (2007), albeit at a smaller subcatchment scale. In this approach, site-based assessments of habitat and several biotic parameters are calculated and then aggregated to the reach scale. On the basis of their representativeness in a subcatchment along with their condition, a set of priority reaches can be targeted for conservation.

RIVER ASSESSMENTS AND THE IMPORTANCE OF THE FUNCTIONAL PROCESS ZONE SCALE

National surveys of rivers have been undertaken in Australia (Norris *et al.*, 2007), Sweden (Wiederholm and Johnson, 1997), the United Kingdom (Wright, 1995; Raven *et al.*, 1998b), and United States (Hughes *et al.*, 2000). Many of these programs have not, however, explicitly recognized or effectively incorporated important concepts of scale and hierarchy. Data are commonly collected at a site and then aggregated to the catchment level. The assessment of river condition for the Murray–Darling Basin (Australia) is a good example. This basin drains an area of 1.07 million km^2 and contains more than 78,000 km of river channels (when measured at the 1:250,000 scale). For this assessment, physical and biological data were collected at 3457 sites. At the request of the management agency, these data were aggregated to the reach scale, which had an average length of 22 km (see Norris *et al.*, 2007 for methods to define reaches). This approach ignored those key principles that underpin hierarchy theory, such that the features of any level within a hierarchically organized system are a composite of both top-down and bottom-up influences. In particular, it is the levels immediately above and below that have the greatest influence on the character of the level of focus. Thus, many assessments of river condition use data collected at an inappropriately low level or scale (e.g., the site scale) to infer catchment-scale condition.

With the increasing emphasis of catchment-based assessments of river systems, it is vital to collect, collate, and present information at the correct scale. The information required by natural resource management agencies to undertake condition assessments of rivers at the catchment scale provides an important spatial domain or boundary for reporting. Within this context and given the framework of the RES, we suggest that information should be *expressed at the scale of the FPZ.*

The utility of this scale is illustrated by the application of FPZs to the auditing and assessment of rivers in the Murray–Darling Basin, a large, highly exploited riverine landscape in southeastern Australia. The preliminary Sustainable Rivers Audit of this basin in 2001 employed FPZs to stratify data collection of biological information (fish, macro invertebrates, and vegetation). This approach may also be used to assist in the development of indicators for suitable rehabilitation (Thoms *et al.*, 2007). The Sustainable Rivers Audit is a program to systematically assess the river health across the Murray–Darling Basin by collecting information in a consistent way across a range of indicators. The Audit reports on the condition of the basin's rivers in catchments for which indicators are collected from FPZs. The collection of information on the distribution and character of FPZs within the Murray–Darling also allowed for the assessment of the physical condition of rivers. These physical data, in turn, provided insights, albeit indirectly, about the biological status of the river network. The common link between the biological and physical components of a river system is the expression of physical habitat as a template for biological communities (Southwood, 1977). As noted previously, from a biological perspective, physical habitat is considered as a template upon which ecological organization and dynamics of ecosystems are observed (Townsend and Hildrew, 1994). Audit protocols require a comparison of the physical, chemical, and biological components from similar river types, or in this case from FPZs. In addition, physical diversity (a measure of the abundance and frequency of different FPZs) is also a useful indicator of the condition of river systems (Thoms *et al.*, 2004).

In the assessment of the physical condition of rivers within the Murray–Darling Basin, Thoms *et al.* (2007) used a number of factors that contribute to the physical condition of rivers to form a single index of condition. Here, the aggregated index was composed of four subindices that consider the influence of changes in habitat, hydrology, land use, and sediment transport on the physical condition of rivers (see Norris *et al.*, 2007 and Thoms *et al.*, 2007, for a full description of methods). Briefly, the physical condition of the rivers was measured via an aggregate Environment Index (ARC_e), which is composed of four unweighted subindices (catchment disturbance or *CDI*; habitat or *HI*; hydrological disturbance or *HDI*; and suspended sediment and nutrients or *SSNLI*). Each index is scaled between 0 and 1, with a value of 1 corresponding to pristine conditions while 0 indicates the most degraded condition. The range of scores for the aggregate ARC_e was divided into four categories: (1) 1.0–0.75 = largely unmodified; (2) 0.75–0.5 = moderately modified; (3) 0.5–0.25 = substantially modified; and (4) 0.25–0 = severely modified. The four unweighted subindices were combined, using the standardized Euclidean distance procedure to calculate the ARC_e (Eq. 8.1). The overall Environmental Index (ARC_e) was employed for general assessment, and the subindices were used for more specific interpretation of the physical condition.

$$ARC_e = 1 - \frac{\sqrt{\left[(1-CDI)^2 + (1-HI)^2 + (1-HD)^2 + (1-SSNLI)^2\right]}}{\sqrt{n}} \tag{8.1}$$

The Murray–Darling system drains the inland slopes of the southeastern highlands of Australia. The hydrogeologic evolution of the drainage network of the basin, and hence the diversity of contemporary river systems, is strongly associated with the development of the

(a)

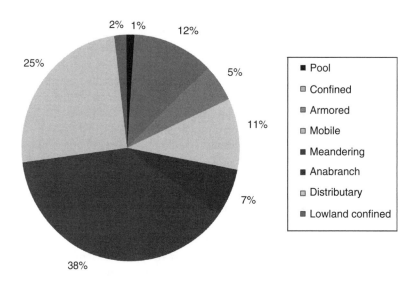

(b)

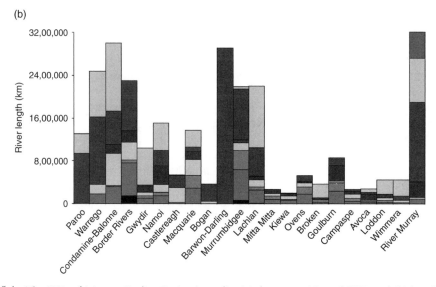

FIGURE 8.1   The FPZs of Murray–Darling Basin, Australia: (a) the composition of FPZs and (b) lengths of FPZs within the various subcatchments of the Murray–Darling Basin. Modified from Thoms *et al.* (2007). (See color plate 10).

landscape of southeastern Australia. Eight FPZs have been identified in the basin; a description of each is given in Chapter 4. Briefly, each FPZ has a distinct valley floor trough–river association in terms of the degree of valley dimensions, gradient, stream power (or energy), boundary material or sediment yield, and river channel planform and cross-sectional character. The different assemblages of morphological units in the channel and on the floodplain influence the presence and structure of in-stream habitat and associated biological communities (Thoms *et al.*, 2004; Boys and Thoms, 2006). The anabranch FPZ is the dominant FPZ in the Murray– Darling (see Fig. 8.1), comprising 39% of the total length of rivers mapped (2.8 million km). This is followed by the distributary, confined, mobile, and meandering FPZs, which account for 25, 11.5, 10, and 7.5% of the total river length, respectively. The pool FPZ was the rarest FPZ, accounting for less than 1% of the total river length.

None of the sub-catchments in the Murray–Darling Basin contains all FPZs (see Fig. 8.1). The Murray, Goulburn, Lachlan, Macquarie, and Murrumbidgee Rivers all include seven of the eight FPZs identified, whereas the Bogan and Darling Rivers each have only two FPZs. As a result, there is a great variation in the composition of FPZs between the different subcatchments of this basin. In the Barwon–Darling, for example, only two FPZs (anabranch and distributary FPZs) occur, and the anabranch FPZ accounts for 99% of its total river length. By comparison, the Campaspe River in Northern Victoria, Australia, contains five FPZs: confined, armored, mobile, meandering, and anabranch, and these account for 18.1, 20.5, 26.7, 21.8, and 12.9% of the total length of this river, respectively. Differences in the physical diversity of the subcatchments in terms of the length of various FPZs can be expressed using Simpson Diversity Index (D_{si}) (Magurran, 2004). Diversity values range from 0.01 for the Barwon–Darling to 0.82 for the Macquarie River (Fig. 8.2). On an average, these subcatchments in the Murray Basin have a higher physical diversity of FPZs compared to an index for the Darling Basin. However, these rivers in the Murray Basin have a lower overall variation in physical diversity compared to the Darling when calculated at the basin scale. The mean and range of Simpson Diversity Index values for the Murray Basin are 0.676 and 0.500–0.780, respectively, compared to 0.586 and 0.199–0.820 for the Darling Basin.

The majority of river lengths within the Murray–Darling Basin had an ARC_e classification of moderately modified (66.4%), while 20% of the total length of rivers was classified as substantially modified. Notable differences were also present in FPZs. All FPZs within the Murray–Darling Basin consisted primarily of river lengths classified as moderately modified (Fig. 8.3). However, the anabranch, confined, distributary, low-confined, meandering, and mobile FPZs have greater than 50% of their river length classified as moderately modified, while only the armored and confined FPZs have a significant proportion of their length as largely unmodified (~16 and 17%, respectively). Many FPZs, however, have large percentages of their reaches classified as substantially modified, most notably the armored (~32%),

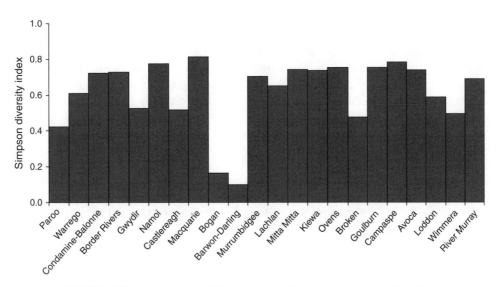

FIGURE 8.2 The diversity of FPZs within the Murray–Darling Basin, Australia.

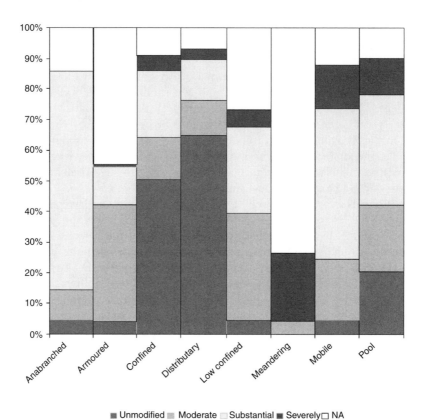

FIGURE 8.3 The physical assessment of functional process zones within the Murray Darling Basin, Australia. From Thoms et al. (2007).

confined (~23%), meandering (~25%), mobile (~30%), and pool FPZs (~42%), with both the anabranch and distributary FPZs also having greater than 10% of their river length classified as substantially modified. Thus, in terms of their overall condition, most FPZs in the Murray–Darling Basin have considerable reach lengths classified as either moderately modified or substantially modified.

Significant sections of the river network in the Murray–Darling Basin are degraded, most of which are in lowland sections of the basin. Such areas in the basin are dominated by anabranch, distributary, and meandering FPZs. However, current management activities, such the First River Health Assessment, have a bias toward the higher-energy upland areas, which have distinct FPZs. This poses inherent problems because it can be difficult to apply broad principles derived from upland areas to proposed management activities in lowland areas (Thoms and Sheldon, 2000). Functional process zones differ in structure and function and can, therefore, respond to natural and human-induced disturbance in disparate ways. Some responses are subtle and within the natural range of river system functions, while others are more obvious and result in major changes to the character and functioning of the river. These responses can impinge upon the health or condition of the river system. The river characterization exercise undertaken in this Australian study underpins the current, more detailed auditing of the condition of river systems in the Murray–Darling Basin. However, many large rivers around the world have been studied to only a limited degree, and insufficient data are available to make responsible decisions. This is especially problematic when the river networks are markedly unpredictable in the flow of water and movement of sediment. It is the responsibility of government to ensure that appropriate data are available on which to base responsible decisions. The widely accepted *precautionary principle* supports this recommendation. For example, extractions of water, sand, and gravel should be delayed until there is sufficient understanding to enable the stakeholders of the river system to make reasonable predictions about their impacts.

DETERMINING ENVIRONMENTAL WATER ALLOCATIONS

Allocating water to sustain natural ecosystems, restore rivers degraded by over abstraction, and protect biodiversity has become a key issue in river management (Thoms and Sheldon, 2000). One of the goals of environmental flow management is to allocate water to maintain riverine habitats (e.g., "PHABSIM", Gore and Nestler, 1988, and "Tennant (Montana) Method", Tennant, 1976). Although the definition of habitat varies within the disciplines of hydrology, fluvial geomorphology, and ecology, it should always be defined both with reference to the species being considered and in terms of physical and biological properties. As such, habitat is interdisciplinary, rather than discipline-specific. Habitat also sits within a hierarchical context, where biotic and abiotic processes that shape habitats occur at multiple spatial and temporal scales. Thus, maintenance of habitat as an endpoint in environmental flow methods is meaningless without reference to: (i) an ecological entity; (ii) the hierarchical organization of river systems; and (iii) knowledge of the intended scale for future river management.

Environmental flow management is frequently concerned with the question: "How much water do we need to allocate to protect and conserve river function?" Outside of an interdisciplinary framework, this question is likely to have three answers because hydrologists, geomorphologists, and ecologists view river systems from the experience of their own disciplines. For example, from a geomorphological perspective, water allocations are required to maintain the structure and function of natural physical features of the river channel (Gippel and Stewardson, 1998). From a biological perspective, water allocations are required to maintain the structure and functioning of populations, communities, and ecosystems. Hence, environmental water allocations generated outside of an interdisciplinary approach may never fully protect and conserve river function, because they do not consider all components of river system and are not cognizant of multiscale linkages among disciplines. The need to consider entire catchments has emerged as a major principle of water management over the last decade as a response to past mistakes. All sources and sinks of water and other materials are considered. Water diverted from rivers impacts not only on surface processes but also on groundwater systems. Environmental flow management requires a comprehensive spatial analysis and identification of an appropriate time base for analysis.

Many different approaches and techniques are available to assist in determining environmental water allocations. A recent review by Tharme (2003) of 207 individual methods noted that these generally occurred at two levels: (1) a reconnaissance-level approach that relied on hydrological methods, such as the Tennant method (Tennant, 1976) and (2) a set of more comprehensive methods of assessment that relied on either hydraulic simulation, such as the instream flow incremental methodology (IFIM), or *holistic* methods. Overall, the methods reviewed by Tharme (2003) are reductionist in approach and use data collected at spatially constrained sites or reaches to build an environmental flow regime. Reliance on methods like IFIM (Bovee and Milhous, 1978) and the building block methodology (King and Louw, 1998), though extremely useful in the initial period of environmental water allocations, fails to recognize the limitation of bottom-up influences in hierarchical organizations such as river ecosystems (Thoms and Parsons, 2002). The influence of lower levels of organization within complex systems gets progressively weaker at successively higher levels, thereby limiting the domain of influence. Advances in understanding environmental water allocations and management, therefore, require bottom-up and top-down approaches that capture a continuum of hierarchical influences.

Using the concept of hierarchy within the context of environmental water allocations, top-down constraints must be recognized. Dollar *et al.* (2007) suggested that if one were to employ a conservation ecology analogy, a top-down approach would recognize the character of the riverine landscape and its hydrology at different scales. Here the management objective would be to maintain the diversity or heterogeneity of this landscape. Managing landscape diversity or heterogeneity is an essential component in conserving system resilience (Pickett *et al.*, 2003) and, in the context of environmental flows, the resilience of riverine ecosystems.

Currently in Australia and elsewhere, environmental flow strategies most often view rivers as uniform and fail to consider spatial and temporal complexity within the riverine landscape. A recent study by Thoms *et al.* (2005) demonstrated a complex spatial pattern of hydrological character in a large Australian dryland river system. This study of the Macintyre River identified six distinct hydrological zones along the river, using multivariate statistics. Full details of the methods employed are given in Thoms *et al.* (2005) and Thoms and Parsons (2003). These hydrological zones represent patches within the hydrological landscape mosaic of the Macintyre, and they correspond to the main FPZs of that river (Thoms *et al.*, 2005). This hydrological study also noted that the dominant timescale of each differed between the river zones. The hydrological character of the headwater zones in the Macintyre was characterized by short-term variables corresponding to individual floods, whereas longer temporal variables, characteristic of event sequencing, better represented these zones lower in catchment. Thus, the spatial and temporal complexity identified at this larger scale requires environmental water allocations to be managed at scales that capture the appropriate patterns of hydrological character in the system in question.

The spatial and temporal complexity of hydrological character within the Macintyre River is an example of a heterogeneous hydrological character of a riverine landscape. Recognition of hydrological mosaics has several implications for environmental flow strategies. The timescale of flow variables associated with the spatial arrangement of the different hydrological patches needs to be recognized so that management intervention can be placed at the appropriate spatial and timescale. Different targets of flow restoration need to be set for individual hydrological zones whereby the attributes of flow must be manipulated, restored, or conserved in accordance with the different timescales of hydrological influence. Maintaining the hydrological integrity of individual zones would allow maintenance of the diversity of the broader mosaic of the hydrological landscape within a catchment. Environmental water allocations are produced through manipulation of the hydrological regime. At what scale should these hydrological manipulations be made to predict physical and biological responses? At a particle scale, flow hydraulics influences the character of the riverbed substratum (Lancaster and Belyea, 1997); and if macroinvertebrates are the diagnostic fauna, the corresponding level of biological organization may be that of an individual organism. At a larger scale, the frequency of a flow partly determines the morphology of river zones (e.g., FPZs), and the corresponding level of biological organization is that of a macroinvertebrate community.

In many rivers, macroinvertebrate communities are evaluated at the local scale and then used as primary biological indicators in environmental flow assessments. These community-level attributes, however, may be inappropriate because of the inherent spatial and temporal complexity in hydrological and geomorphological character. For example, given the dominance of short-term, pulse-scale hydrological variables in the headwater zones of the Macintyre River, it would be more appropriate to monitor populations of individual organisms at small patches within a reach (cf., Dollar *et al.*, 2007). In those zones located further downstream, characterized hydrologically by events over longer timescales, community-level attributes could be monitored. Thus, biological indicators used to monitor environmental flows must match the appropriate scales of physical and hydrological processes that occur in the river system. Incorporation of this multi-dimensional spatial and temporal approach into existing environmental flow strategies will advance the application of the natural flow paradigm (Poff *et al.*, 1997) and, by association, may improve ecosystem responses to managed flows (Thoms and Parsons, 2003).

Employing a hierarchical approach, as advocated by the use of FPZs, places the collective use of top-down and bottom-up methods in a multidimensional context and facilitates an adaptive management approach for the setting and management of environmental water alloca-tions (Dollar *et al.*, 2007). The management and conservation of diversity in the hydrological landscape is a higher-level objective within which methods like IFIM and the building block methodology could be used to assess flow needs in specific patches of the mosaic. Determining important biophysical flows then has a context. That is, important biophysical flows, whether applied to a single species at a site or instead to a range of organisms within multiple river patches, become the focus for hypothesis-based monitoring. Monitoring through testing hypotheses about the functions of certain flows and the implications of changing these flows in river ecosystems overcomes the much-critiqued approach of monitoring for the sake of monitoring.

SUMMARY

The importance of river classification in conservation planning and management is widely recognized. The river characterization approach presented as part of the RES provides an excellent framework to classify riverine landscapes at various scales. By focusing on the scale of FPZs (which are large tracts of river that contain a similar physical structure and experience similar physical processes), scientists, river managers, and people interested in river rehabilita-tion will be operating at a scale that is appropriate for entire river networks. The identification and mapping of FPZs can provide a framework for conservation planning and prioritization of management activities within river networks.

Interdisciplinary research in river ecosystems is a relatively young endeavor and one that is fraught with problems – linking across scales and integrating different disciplinary approaches and conceptual tools. Frameworks are useful tools for achieving this because they help define the bounds for the selection and solution of problems. They also indicate the role of empirical assumptions, carry the structural assumptions, show how facts, hypotheses, models, and expec-tations are linked, and indicate the scope at which a generalization or model applies. The RES provides such an integrative framework for the study of entire riverine landscapes. A framework is neither a model nor a theory because models describe how things work and theories explain phenomena. In contrast, a conceptual framework helps to order phenomena and materials, thereby revealing patterns. To advance interdisciplinary research on riverine landscapes, we require the development and articulation of the RES as a framework in order to unify the field of river science and ensure interdisciplinary interaction at appropriate scales.

Concluding Remarks

In the preface of this book, we outlined our goals and included caveats about various riverine theories, including the RES. We want to again emphasize that theories are, in a sense, formed of unfired clay and require extensive shaping and remolding before they accurately model the real world (if ever). As scientists, we too often forget that paradigms, theories, and models are merely aggregations of assumptions and hypotheses – no matter how many disciples they may have attracted. We have always envisioned the RES as a heuristic model, which we hope will grow and develop over time. To improve this continually developing synthesis, we need to understand why aspects and predictions of the RES seem true in one riverine ecosystem but are not verified to the same degree, if at all, in another.

The RES was designed to serve both as a description of hierarchical patterns and processes at multiple dimensions of riverine landscapes from headwaters to great rivers and as a framework for research at multiple spatiotemporal scales. Its original 2006 publication in *River Research and Applications* focused on pristine ecosystems, but we have provided here some tools for tackling real-world challenges of river monitoring, assessment, management, rehabilitation, and conservation. The applications of the RES to these tasks still need considerable refinement, but we hope the examples of methods and approaches we have provided here will help in the task of making lotic environments around the world more healthy. As of January 2008, components of the RES are already being used in river management in portions of Australia, South Africa and are being considered for use by governments and NGOs in other countries, including the United States.

We encourage you to keep an open but skeptical mind about the RES and to send us your comments, suggestions, and criticisms on any material in this book. In particular, we would like your suggestions for changes or additions to the model tenets. To that end, we have included a separate page for your suggestions on one or more new tenets. Feel free to copy this page and send your suggestions to one or all of us. If we use your ideas in a new journal article, or write a second edition to this book, or even give a seminar about the RES, we will definitely acknowledge your contributions. We could also benefit from suggestions on applications of the RES to altered riverine landscapes.

We hope that this book has sparked your interest in studying broad-scale and hierarchical patterns in riverine ecosystems and that you will continue to evaluate carefully hypotheses within this synthesis as well as concepts developed by other scientists. Whether you are a student or a professional scientist, feel free to corner us at some science conference and express your ideas, even if we have never been introduced – conversations over a cool beer are particularly encouraged!

Sincerely,
Jim, Martin, and Mike

Proposal for a New Tenet

Your Name:

Institutional Affiliation:

Contact Information:

Possible Tenet Section:
[i.e., Distribution of Species, Community Regulation, or Ecosystem and Landscape Processes]

Full Tenet Description:

Justification:
[Summarize your argument and provide references if you have them.]

Literature Cited

Aarts, B.G.W., F.W.B. Van den Brink, and P.H. Nienhuis. 2004. Habitat loss as the main cause of the slow recovery of fish faunas of regulated rivers in Europe: The transversal floodplain gradient. River Research and Applications 20:3–23.

Adams, S., M. Warren, and W. Haag. 2004. Spatial and temporal patterns in the fish assemblages of upper coastal plain streams, Mississippi, USA. Hydrobiologia 528:45–61.

Anderson, C. and G. Cabana. 2007. Estimating the trophic position of aquatic consumers in river food webs using stable nitrogen isotopes. Journal of the North American Benthological Society 26:273–285.

Ahn, C., R.E. Sparks, and D.C. White. 2004. A dynamic model to predict responses of millets (*Echiochloa* sp.) to different hydrologic conditions for the Illinois floodplain-river. River Research and Applications 20:485–498.

Akbulut, N. 2005. The determination of relationship between zooplankton and abiotic factors using canonical correspondence analysis (CCA) in the Ova stream (Ankara/Turkey). Acta Hydrochimica et Hydrobiologica 32:434–441.

Albert, R. and A.-L Barabási. 2002. Statistical mechanics of complex networks. Reviews of Modern Physics 74:47–97.

Allen, T.F.H. and T.B. Starr. 1982. Hierarchy: Perspectives for ecological diversity. University of Chicago Press, Chicago, IL.

Amoros, C. and G. Bornette. 2002. Connectivity and biocomplexity in waterbodies of riverine floodplains. Freshwater Biology 47:761–776.

Angradi, T. 1994. Trophic linkages in the lower Colorado River: Multiple stable isotope evidence. Journal of the North American Benthological Society 13:479–495.

Araujo-Lima, C.A.R.M., B.R. Forsberg, R. Victoria, and L.A. Martinelli. 1986. Energy sources for detritivorous fishes in the Amazon. Science 234:1256–1258.

Armitage, P.D., K. Szoszkiewicz, J.H. Blackburn, and I. Nesbitt. 2003. Ditch communities: A major contributor to floodplain biodiversity. Aquatic Conservation: Marine and Freshwater Ecosystems 13:165–185.

Arrington, D.A. and K.O. Winemiller. 2006. Habitat affinity, the seasonal flood pulse, and community assembly in the littoral zone of a Neotropical floodplain river. Journal of the North American Benthological Society 25:126–141.

Arrington, D.A., K.O. Winemiller, and C.A. Layman. 2005. Community assembly at the patch scale in a species rich tropical river. Oecologia 144:157–167.

Arscott, D.B., K. Tockner, and J.V. Ward. 2005. Lateral organization of aquatic invertebrates along the corridor of a braided floodplain river. Journal of the North American Benthological Society 24:934–954.

Aspetsberger, F., F. Huber, S. Kargi, B. Scharinger, P. Peduzzi, and T. Hein. 2002. Particulate organic matter dynamics in a river floodplain system: Impact of hydrological connectivity. Archiv für Hydrobiologie 156:23–42.

Bailey, K.D. 1984. Typologies and taxonomies: An introduction to classification techniques. Sage, Thousand Oaks, CA.

Balcombe, S.R., S.E. Bunn, F.J. McKenzie, and P.M. Davies. 2005. Variability of fish diets between dry and flood periods in an arid zone floodplain river. Journal of Fish Biology 67:1552–1567.

Baldwin, D.S., A.M. Mitchell, G.N. Rees, G.O. Watson, and J.L. Williams. 2005. Nitrogen processing by biofilms along a lowland river continuum. Journal of the North American Benthological Society 22:319–326.

Baptist, M.J., M. Haasnoot, P. Cornelissen, J. Icke, G. van der Wedden, H.J. de Vriend, and G. Gugic. 2006. Flood detention, nature development and water quality along the lowland river Sava, Croatia. Hydrobiologia 565:243–257.

Baranyi, C., T. Hein, C. Holarek, S. Keckeis, and F. Schiemer. 2002. Zooplankton biomass and community structure in a Danube River floodplain system: Effects of hydrology. Freshwater Biology 47:1–10.

Barko, V.A., D.P. Herzog, and M.T. O'Connell. 2006. Response of fishes to floodplain connectivity during and following a 500-year flood event in the unimpounded Upper Mississippi River. Wetlands 26:244–257.

Barrett, G.W., J.D. Peles, and E.P. Odum. 1997. Transcending processes and the levels of organisation concept. Bioscience 47:531–535.

Bayley, P.B. 1989. Aquatic environments in the Amazon Basin, with an analysis of carbon sources, fish production, and yield. Pages 399–408 *in*: D.P. Dodge (ed.), Proceedings of the International Large Rivers Symposium. Canadian Special Publication in Fisheries and Aquatic Sciences, 106.

Belbin, L. 1993. PATN technical reference. CSIRO Division of Wildlife and Ecology, Canberra, Australia.

Belbin, L. and C. McDonald. 1993. Comparing three classification strategies for use in ecology. Journal of Vegetation Science 4:341–348.

Benda, L., L.R. Poff, D. Miller, T. Dunne, G. Reeves, M. Pollock, and G. Pess. 2004. Network dynamics hypothesis: Spatial and temporal organization of physical heterogeneity in rivers. BioScience 54:413–427.

Bendix, J. and C.R. Hupp. 2000. Hydrological and geomorphological impacts on riparian plant communities. Hydrological Processes 14:2977–2990.

Benke, A.C. 2001. Importance of flood regime to invertebrate habitat in an unregulated river-floodplain ecosystem. Journal of the North American Benthological Society 20:225–250.

Bergkamp, G., 1995. A hierarchical approach for desertification assessment. Environmental Monitoring and Assessment 37:59–78.

Biggs, H.C., 2003. Integration of science: Successes, challenges, and the future. Pages 469–487 *in*: J.T. Du Toit, K.H. Rogers, and H.C. Biggs (eds.), The Kruger Experience: Ecology and Management of Savanna Heterogeneity. Island Press, Washington.

Biggs, B.J.F., M.J. Duncan, I.G. Jowett, J.M. Quinn, C.W. Hickey, R.J. Davies-Colley, and M.E. Close. 1990. Ecological characterisation, classification and modelling of New Zealand rivers: An introduction and synthesis. New Zealand Journal of Marine and Freshwater Research 24:277–304.

Biggs, B.J.F. and R.A. Smith. 2002. Taxonomic richness of stream benthic algae: Effects of flood disturbance and nutrients. Limnology and Oceanography 47:1175–1186.

Bini, L.M., S.M. Thomaz, and D.C. Souza. 2001. Species richness and beta-diversity of aquatic macrophytes in the Upper Parana River floodplain. Archiv für Hydrobiologie 151:511–525.

Bird, M.I., P. Giresse, and S. Ngos. 1998. A seasonal cycle in the carbon-isotope composition of organic carbon in the Sanaga River, Cameroon. Limnology and Oceanography 43:143–146.

Bisson, P.A. and D.R. Montgomery. 1996. Valley segments, stream reaches and channel units. Pages 23–52 *in*: F.R. Hauer and G.A. Lamberti (eds.), Methods in Stream Ecology. Academic Press, San Diego.

Bisson, P.A., R.A. Nielsen, and R.A. Palmason. 1982. A system of naming habitat types in small streas, with examples of habita utilization by salmonids during low streamflow. Pages 62–73 *in*: N.B. Armantrout (ed.), Acquisition and Utilization of Aquatic Habitat Inventory Information. Western Divison of the American Fisheries Society, Oregon, USA.

Bluck, B.J. 1982. Texture of gravel bars in braided streams. Pages 339–355 *in*: R.D. Hey, J.C. Bathurst and C.R. Thorne (eds.), Gravel Bed Rivers. Wiley, New York.

Bonada, N., M. Rieradevall, N. Prat, and V.H. Resh. 2006. Benthic macroinvertebrate assemblages and macrohabitat connectivity in Mediterranean-climate streams of northern California. Journal of the North American Benthological Society 25:32–43.

Boon, P.J., B.R.Davies, and G.E.Petts (eds.). 2000. Global Perspectives in River Conservation. Wiley, Chichester, UK.

Bormann, F.H. and G.E. Likens. 1979. Catastrophic disturbances and the steady state in northern hardwood forests. American Scientist 67:660–669.

Bornette, G., C. Amoros, and N. Lamouroux. 1998. Aquatic plant diversity in riverine wetlands: The role of connectivity. Freshwater Biology 39:267–283.

Boulton, A.J. 2003. Parallels and contrasts in the effects of drought on stream macroinvertebrate assemblages. Freshwater Biology 48:1173–1185.

Boulton, A.J., C. Hakenkamp, M. Palmer, and D. Strayer. 2002. Freshwater meiofauna and surface water-sediment linkages: A conceptual framework for cross-system comparisons. Pages 241–259 *in*: S.D. Rundle, A.L. Robertson, and J.M. Schmid-Araya (eds.), Freshwater Meiofauna: Biology and Ecology. Backhuys, Leiden, The Netherlands.

Boulton, A.J. and E.H. Stanley. 1995. Hyporheic processes during flooding and drying in a Sonoran desert stream. II. Faunal dynamics. Archiv für Hydrobiologie 134:27–52.

Bovee, K.D. and R.T. Milhous. 1978. Hydraulic simulation in instream flow studies: Theory and techniques. US Fish and Wildlife Service Biological Services Program, Cooperative Instream Flow Service Group, Instream Flow Information Paper No. 5. FWS/OBS-78-33. 130 p.

Bowen, Z.H., K.D. Bovee, and T.J. Waddle. 2003. Effects of flow regulation on shallow-water habitat dynamics and floodplain connectivity. Transactions of the American Fisheries Society 132:809–823.

Boyero, L. and R.C. Bailey. 2001. Organization of macroinvertebrate communities at a hierarchy of spatial scales in a tropical stream. Hydrobiologia 464:219–225.

Boys, C.A., G. Esslemont, and M.C. Thoms. 2005. Fish habitat assessment and protection in the Barwon-Darling and Paroo Rivers. Technical report to the Department of Agriculture, Fisheries and Forestry Australia., 118 p.

Boys, C.A. and M.C. Thoms. 2006. A hierarchical scale approach to the assessment of fish assemblages and their habitat associations in large dryland rivers. Hydrobiologia 572:11–31.

Brett, M.T. and D.C. Müller-Navarra. 1997. The role of highly unsaturated fatty acids in aquatic food web processes. Freshwater Biology 38:483–499.

Brierley, G. and E.D. Hickin. 1991. Floodplain development based on selective preservation of sediments, Squamish River, British Columbia. Geomorphology 4:381–391.

Brierley, G. and K. Fryirs. 2005. Geomorphology and river management. Blackwell Publishing, Cornwall.

Brown, K.M. 2001. Gastropoda. Pages 297–329 *in*: J.H. Thorp and A.P. Covich (eds.), Ecology and classification of North American freshwater invertebrates. Academic Press (Elsevier), San Diego.

Brown, K.M., J.E. Alexander, and J.H. Thorp. 1998. Differences in the ecology and distribution of lotic pulmonate and prosobranch gastropods. American Malacological Bulletin 14:91–101.

Bruesewitz, D.A., J.L. Tank, M.J. Bernot, W.B. Richardson, and E.A. Strauss. 2006. Seasonal effects of the zebra mussel (*Dreissena polymorpha*) on sediment denitrification rates in Pool 8 of the Upper Mississippi River. Canadian Journal of Fisheries and Aquatic Science 63:957–969.

Brunke, M. 2002. Floodplains of a regulated southern alpine river (Brenno, Switzerland): ecological assessment and conservation options. Aquatic Conservation: Marine and Freshwater Ecosystems 12:583–599.

Bukaveckas, P.A., J.J. Williams, and S.P. Hendricks. 2002. Factors regulating autotrophy and heterotrophy in the main channel and an embayment of a large river reservoir. Aquatic Ecology 36:355–369.

Bunn, S.E. and P.M. Davies. 1999. Aquatic food webs in turbid, arid zone rivers: Preliminary data from Cooper Creek, Queensland. Pages 67–76 *in*: R.T. Kingford (ed.), Free-Flowing River: The Ecology of the Paroo River (Ed. R.T. Kingsford), pp. 67–76. New South Wales National Parks and Wildlife Service, Sydney.

Bunn, S.E., P.M. Davies, and M. Winning. 2003. Sources of organic carbon supporting the food web of an arid zone floodplain river. Freshwater Biology 48:619–635.

Bunn, S.E., M.C. Thoms, S.K. Hamilton, and S.J. Capon. 2006. Flow variability in dryland rivers: Boom, bust and the bits in between. River Research and Application 22:179–186.

Burgherr, P., J.V. Ward, and C.T. Robinson. 2002. Seasonal variation in zoobenthos across habitat gradients in an alpine glacial floodplain (Val Roseg, Swiss Alps). Journal of the North American Benthological Society 21:561–575.

Burns, A. and K.F. Walker. 2000. Effects of water level regulation on algal biofilms in the River Murray, South Australia. Regulated Rivers: Research and Management 16:433–444.

Cabana, G. and J.B. Rasmussen. 1994. Modelling food chain structure and contaminant bioaccumulation using stable nitrogen isotopes. Nature 372:255–257.

Caissie, D. 2006. The thermal regime of rivers: A review. Freshwater Biology 51:1389–1406.

Callum, C., G.J. Brierley, and M.C. Thoms. in press. The spatial organisation of river systems. Pages: xxx–xxx *in*: G.J. Brierley and K.A. Fryirs (eds.), River Futures. Island Press, Washington DC.

Calow, P. 1992. Can ecosystems be healthy? critical consideration of concepts. Journal of Aquatic Ecosystem Health 1:1–5.

Campos, D., J. Fort, and V. Mendez. 2006. Transport on fractal river networks: Application to migration fronts. Theoretical Population Biology 69:88–93.

Caraco, N.F., J.J. Cole, P.A. Raymond, D.L. Strayer, M.L. Pace, S.E.G. Findlay, and D.T. Fischer. 1997. Zebra mussel invasion in a large, turbid river: Phytoplankton response to increased grazing. Ecology 78:588–602.

Cardinale, B.J., M.A. Palmer, C.M. Swain, S. Brooks, and N.L. Poff. 2002. The influence of substrate heterogeneity on biofilm metabolism in a stream ecosystem. Ecology 83:412–422.

Carini, G., J.M. Hughes, and S.E. Bunn. 2006. The role of waterholes as "refugia" in sustaining genetic diversity and variation of two freshwater species in dryland river systems (Western Queensland, Australia). Freshwater Biology 51:1434–1446.

Carpenter, K.E. 1928. Life in inland waters. Sidgwick and Jackson, London, U.K.

Carvalho, P., S. Thomaz, and L. Bini. 2003. Effects of water level, abiotic, and biotic factors on bacterioplankton abundance in lagoons of a tropical floodplain (Parana River, Brazil). Hydrobiologia 510:67–74.

Casas, J.J., C. Zamora-Munoz, F. Archila, and J. Alba-Tercedor. 2000. The effect of a headwaterdam on the use of leaf bags by invertebrate communities. Regulated Rivers: Research and Management 16:577–591.

Casper, A.F. and J.H. Thorp. 2007. Diel and lateral patterns of zooplankton distribution in the St. Lawrence River. River Research and Applications 23:73–86.

Chen, Z., L. Yu, and A. Gupta (eds.) The Yangtze River. A. GuptaSpecial Issue of Geomorphology. 41,(2–3):73–248.

Chessman, B., I. Growns, J. Currey, and N. Plunkett-Cole. 1999. Predicting diatom communities at the genus level for the rapid biological assessment of rivers. Freshwater Biology 41:317–331.

Church, M. 1992. Channel morphology and typology. Pages 126–143 *in*: P. Calows and G.E. Petts (eds.), The rivers handbook. Blackwell, London, UK.

Church, M. 2002. Geomorphic thresholds in riverine landscapes. Freshwater Biology 47:541–557.

Clements, F.E. 1916. Plant Succession: Analysis of the Development of Vegetation. Carnegie Institute of Washington Publication, No. 42, Washington, DC, USA.

Coates, D.R. 1984. Geomorphology and public policy. Pages 96–115 *in*: J.E. Costa and P.J. Fleisher (eds.), Developments and Applications of Geomorphology. Springler Verlag, Berlin.

Cohen, J.E. and C.M. Newman. 1988. Dynamic basis of food web organization. Ecology 69:1655–1664.

Cohen, R.R.H., P.V. Dresler, E.J.P. Phillips, and R.L. Cory. 1984. The effect of the Asiatic clam, *Corbicula fluminea*, on phytoplankton of the Potomac river, Maryland. Limnology and Oceanography 29:170–180.

Cole, J.J. and N.A. Caraco. 2001. Carbon in catchments: Connecting terrestrial carbon losses with aquatic metabolism. Journal of Marine and Freshwater Research 52:101–110.

Connell, J.H. 1978. Diversity in tropical rainforests and coral reefs. Science 199:1302–1310.

Connell, J.H. and R.O. Slatyer. 1977. Mechanisms of succession in natural communities and their role in community stability and organization. American Naturalist 111:1119–1144.

Cordes, L.D., F.M.R. Hughes, and M. Getty. 1997. Factors affecting the regeneration and distribution of riparian woodlands along a northern prairie river: The Red Deer River, Alberta, Canada. Journal of Biogeography 24:675–695.

Cotner, J.B., J.V. Montoya, D.L. Roelke, and K.O. Winemiller. 2006. Seasonally variable riverine production in the Venezuelan llanos. Journal of the North American Benthological Society 25:171–184.

Cullen, P. 1990. The turbulent boundary between science and water management. Freshwater Biology 24:201–209.

Cupp, C.E. 1989. Valley segment type classification for forested lands of Washington. Timber, Fish, and Wildlife Ambient Monitoring Program, Report TFW-AM-89-001.

Dahm, C.N., M.A. Baker, D.I. Moore, and J.R. Thibault. 2003. Coupled biogeochemical and hydrological responses of streams and rivers to drought. Freshwater Biology 48:1219–1231.

Davies, B.R., R.D. Beilfuss, and M.C. Thoms. 2000. Cahora Bassa retrospective, 1974–1997: effects of flow regulation on the Lower Zambezi River. Verhandlungen Internationale Vereinigung für theoretische und angewandte Limnologie 27:2149–2157.

Davies, B.R. and K.F. Walker. 1986. The ecology of world rivers, Junk, Dordrecht.

Davies, P.E., J. Long, M.B. Brown, H. Dunn, D. Heffer, and R. Knight. 2004. The Tasmanian Conservation of Freshwater Ecosystem Values (CFEV) Framework: Developing a Conservation and Management System for Rivers. Freshwater Protected Areas: New and Existing Tools for Conserving Freshwater Ecosystems in Australia. Environment Australia, Sydney.

Davis, W.M. (1899). The geographic cycle. Geographical Journal 14:481–504.

Dayton, P.K. 1971. Competition, disturbance, and community organization: The provision and subsequent utilization of space in a rocky intertidal community. Ecological Monographs 41:351–389.

Death, R.G. 2002. Predicting invertebrate diversity from disturbance regimes in forest streams. Oikos 97:18–30.

De Bartolo, S.G., S. Gabriele, and R. Gaudio. 2000. Multifractal behaviour of river networks. Hydrology and Earth System Sciences 4:105–112.

de Boer, D.H. 1992. Hierarchies and spatial scale in process geomorphology: A review. Geomorphology 4:303–318.

Deiller, A.F., J.M.N. Walter, and M. Tremolieres. 2001. Effects of flood interruption on species richness, diversity and floristic composition of woody regeneration in the upper Rhine alluvial hardwood forest. Regulated Rivers: Research and Management 17:393–405.

Delong, M.D. 2005. Upper Mississippi River Basin. Pages 327–373 *in*: A.C. Benke and C.E. Cushing (eds.), Rivers of North America. Elsevier, New York, USA.

Delong, M.D. and M.A. Brusven. 1991. Classification and spatial mapping of riparian habitat with applications toward management of streams impacted by nonpoint source pollution. Environmental Management 15:565–571.

Delong, M.D. and M.A. Brusven. 1992. Patterns of periphyton chlorophyll *a* in an agricultural nonpoint source impacted stream. Water Resources Bulletin 28:731–741.

Delong, M.D. and M.A. Brusven. 1998. Macroinvertebrate community structure along the longitudinal gradient of an agriculturally impacted stream. Environmental Management 22:445–457.

Delong, M.D. and N.D. Mundahl. 1995. Habitat availability and invertebrate habitat utilization in the Lower Chippewa River: A river subjected to hydropower peaking. Final Report to the Wisconsin Department of Natural Resources, Eau Claire, WI.

Delong, M.D. and J.H. Thorp. 2006. Significance of instream autotrophs in trophic dynamics of the Upper Mississippi River. Oecologia 147:76–85.

Delong, M.D., J.H. Thorp, K.S. Greenwood, and M.C. Miller. 2001. Responses of consumers and food resources to a high magnitude, unpredicted flood in the upper Mississippi River basin. Regulated Rivers: Research and Management 17:217–232.

Denslow, J.S. 1985. Disturbance-mediated coexistence of species. Pages 307–323 *in*: S.T.A. Pickett and P.S. White (eds.), The Ecology of Natural Disturbance and Patch Dynamics. Academic Press, New York. 472 p.

Dent, C.L., N.B. Grimm, and S.G. Fisher. 2001. Multiscale effects of surface-subsurface exchange on stream water nutrient concentrations. Journal of the North American Benthological Society 20:162–181.

Desgranges, J.L. and B. Jobin. 2003. Knowing, mapping and understanding St. Lawrence biodiversity, with special emphasis on bird assemblages. Environmental Monitoring and Assessment 88:177–192.

Dodds, W.K., K. Gido, M.R. Whiles, K.M. Fritz, and M.J. Matthews. 2004. Life on the edge: The ecology of Great Plains prairie streams. BioScience 54:207–218.

Dodds, P.S. and D.H. Rothman. 2000. Scaling, universality, and geomorphology. Annual Review of Earth and Planetary Sciences 28:571–610.

Dodson, S.I. and D.G. Frey. 2001. Cladocera and other Branchiopoda. Pages 850–914 *in*: J.H. Thorp and A.P. Covich (eds.), Ecology and classification of North American freshwater invertebrates. Academic Press (Elsevier), San Diego.

Dollar, E.S.J., C.S. James, K.H. Rogers, and M.C. Thoms. 2007. A framework for interdisciplinary understanding of rivers as ecosystems. Geomorphology 89:147–162.

Downes, B.J. 1990. Patch dynamics and mobility of fauna in streams and other habitats. Oikos 59:411–413.

Doyle, M.W. 2006. A heuristic model for potential geomorphic influences on trophic interactions in streams. Geomorphology 77:235–248.

Dunham, J.B. and B.E. Rieman. 1999. Metapopulation structure of bull trout: Influences of physical, biotic, and geometrical landscape characteristics. Ecological Applications 9:642–655.

Dunham, J.B., A.E. Rosenberger, C.H. Luce, and B.E. Rieman. 2007. Influences of wildfire and channel reorganization on spatial and temporal variation in stream temperature and the distribution of fish and amphibians. Ecosystems 10:335–346.

Du Toit, J.T., K.H. Rogers, and H.C. Biggs (eds.). 2003. The Kruger Experience: Ecology and Management of Saravana Heterogeneity. Island Press, Washington. 519p.

Eckblad, J.W., C.S. Volden, and L.S. Weilgart. 1984. Allochthonous drift to the main channel of the Mississippi River. American Midland Naturalist 111:16–22.

England, L.E. and A.D. Rosemond. 2004. Small reductions in forest cover weaken terrestrial-aquatic linkages in headwater streams. Freshwater Biology 49:721–734.

Finlay, J.C. 2001. Stable carbon isotope ratios of river biota: Implications for energy flow in lotic food webs. Ecology 84:1052–1064.

Fisher, S.G. 1983. Succession in streams. Pages 7–27 *in*: J.R. Barnes and G.W. Minshall (eds.), Stream Ecology: Application and Testing of General Ecological Theory. Plenum Press, New York.

Fisher, S.G. 1993. Pattern, process, and scale in freshwater systems: Some unifying thoughts. Pages 575–597 *in*: P.S. Giller, A.G. Hildrew, and D.G. Raffaelli (eds.), Aquatic Ecology: Scale, Pattern, and Process. Blackwell Scientific, Oxford.

Fisher, S.G. 1997. Creativity, idea generation, and the functional morphology of streams. Journal of the North American Benthological Society 16:305–318.

Fisher, S.G., L.J. Gray, N.B. Grimm, and D.E. Busch. 1982. Temporal succession in a desert stream ecosystem following flash flooding. Ecological Monographs 52:93–110.

Fisher, S.G., N.B. Grimm, E. Martí, and R. Gómez. 1998. Hierarchy, spatial configuration, and nutrient cycling in a desert stream. Australian Journal of Ecology 23:41–52.

Fisher, S.G., J.B. Heffernan, R.A. Sponseller, and J.R. Welter. 2006. Functional ecomorphology: Feedbacks between form and function in fluvial landscape ecosystems. Geomorphology 89:84–96.

Flory, E.A. and A.M. Milner. 2000. Macroinvertebrate community succession in Wolf Point Creek, Glacier Bay National Park, Alaska. Freshwater Biology 44:465–480.

Forman, R.T.T. and M. Godron. 1986. Landscape ecology. John Wiley and Sons, New York.

Forman, R.T.T. 1997. Landscape mosaics: The ecology of landscapes and regions. Cambridge, London.

Forrest, J. and N.R. Clark. 1989. Characterising grain size distributions: Evaluation of a new approach using a multivariate extension of entropy analysis. Sedimentology 36:711–722.

Forsberg, B.R., C.A.R.M. Araujo-Lima, L.A. Martinelli, R.L. Victoria, and J.A. Bonassi. 1993. Autotrophic carbon sources for fish of the Central Amazon. Ecology 74:643–652.

Fremling, C.R., J.L. Rasmussen, R.E. Sparks, S.P. Cobb, C.F. Bryan, and T.O. Claflin. 1989. Mississippi River fisheries: A case history. Pages 309–351 *in*: D.P. Dodge (ed.), Proceedings of the International Large River Symposium (LARS). Canadian Special Publication of Fishers and Aquatic Science, 106.

Friedman, J.M. and V.J. Lee. 2002. Extreme floods, channel change, and riparian forests along ephemeral streams. Ecological Monographs 72:409–425.

Frissell, C.A., W.J. Liss, C.E. Warren, and M.D. Hurley. 1986. A hierarchical framework for stream habitat classification: Viewing streams in a watershed context. Environmental Management 10:199–214.

Fry, B. 2002. Stable isotopic indicators of habitat use by Mississippi River fish. Journal of the North American Benthological Society 21:676–685.

Fry, B. 2006. Stable Isotope Ecology. Springer, New York, NY. 308 p.

Fry, B. and Y.C. Allen. 2003. Stable isotopes in zebra mussels as bioindicators of river-watershed linkages. River Research and Applications 19:683–696.

Funk, J.L. and J.W. Robinson. 1974. Changes in the channel of the Lower Missouri River and effects on fish and wildlife. Missouri Department of Conservation, Aquatic Series 11, Columbia, Missouri, USA.

Galat, D.L., C.R. Berry, W.M. Gardner, J.C. Hendrickson, G.E. Mestl, G.J. Power, C. Stone, and M.R. Winston. 2005. Spatiotemporal patterns and changes in Missouri River fishes. Pages 249–291 *in*: J.N. Rinne, R.M. Hughes, and B. Calamusso (eds.), Historical changes in large river fish assemblages of the Americas, American Fisheries Society Symposium 45, Bethesda, Maryland, USA.

Galat, D.L., L.H. Frederickson, D.D. Humburg, K.J. Bataille, J.R. Bodie, J. Dohernwood, G.T. Gelwicks, J.E. Havel, D.L. Helmers, J.B. Hooker, J.R. Jones, M.F. Knowlton, J. Kubisiak, J. Mazourek, A.C. McColpin, R.B. Renken, and R.D. Semlitsch. 1998. Flooding to restore connectivity of regulated, large-river wetlands. Bioscience 48:721–733.

Galat, D.L. and I. Zweimüller. 2001. Conserving large-river fishes: Is the highway analogy an appropriate paradigm. Journal of the North American Benthological Society 20:266–279.

Gardiner, V. 1995. Channel networks; progress in the study of spatial and temporal variations of drainage density. Pages 65–86 *in*: A.D. Gurnell and G.E. Petts (eds.), Changing River Channels. Wiley, Chichester, UK.

Georgian, T. and J.H. Thorp. 1992. Effects of microhabitat selection on feeding rates of net-spinning caddisfly larvae. Ecology 73:229–240.

Gerritsen, J. 1995. Additive biological indices for resource management. Journal of the North American Benthological Society 14:451–457.

Gibert, J., D.L. Danielopol, and J.A. Stanford. 1994. Groundwater Ecology. Academic Press, San Diego. 571 p.

Giller, P.S. and B. Malmqvist. 1998. The Biology of Streams and Rivers. Oxford University Press, Oxford, UK.

Gillette, D.P., J.S. Tiemann, D.R. Edds, and M.L. Wildhaber. 2005. Spatiotemporal patterns of fish assemblage structure in a river impounded by low-head dams. Copeia 2005:539–549.

Gippel, G.J. and M.J. Stewardson. 1998. Use of wetted perimeter in defining minimum environmental flows. Regulated Rivers: Research and Management 14:53–67.

Gomi, T., R.C. Sidle, and J.S. Richardson. 2002. Understanding processes and downstream linkages of headwater streams. BioScience 52:905–916.

Goodson, J.M., I.P. Morrissey, P.G. Angold, and A.M. Gurnell. 2000. Seed bank secrets of the riverbank. Vegetation Management in Changing Landscapes, Aspects of Applied Biology 58:205–212.

Gore, J.A. and R.M. Bryant, Jr. 1986. Changes in fish and benthic macroinvertebrate assemblages along the impounded Arkansas River. Journal of Freshwater Ecology 3:333–345.

Gore, J.A. and J.M. Nestler. 1988. Instream flow studies in perspective. Regulated Rivers: Research and Management 2:93–101.

Gosz, J.R. 1991. Fundamental ecological characteristics of landscape boundaries. Pages 52–75 *in*: M.M. Holland, P.G. Risser, and R.J. Naiman (eds.), Ecotones: The role of landscape boundaries in the management and restoration of changing environments. Chapman and Hall, London.

Gotelli, N.J. and C.M. Taylor. 1999. Testing metapopulation models with stream-fish assemblages. Evolutionary Ecology Research 1:835–845.

Greenwood, K.S. and J.H. Thorp. 2001. Aspects of ecology and conservation of sympatric, prosobranch snails in a large river. Hydrobiologia 455:229–236.

Gregory, K.J. 2006. The human role in chaning river channels. Geomorphology 79:172–191.

Grimm, N.B. 1993. Disturbance, succession and ecosystem processes in streams: A case study from the desert. Pages 93–112 *in*: P.S. Giller, A.G. Hildrew, and D.G. Raffaelli (eds.), Aquatic Ecology: Scale, Pattern and Process. Blackwell Scientific Publications.

Grimm, N.B. and S.G. Fisher. 1989. Stability of periphyton and macroinvertebrates to disturbance by flash floods in a desert stream. Journal of the North American Benthological Society 9:293–307.

Groffman, P.M., N.L. Law, K.T. Bell, L.E. Band, and G.T. Fisher. 2004. Nitrogen fluxes and retention in urban watershed ecosystems. Ecosystems 7:393–403.

Grubbs, S. and J. Taylor. 2004. The influence of flow impoundment and river regulation on the distribution of riverine benthic macroinvertebrates at Mammoth Cave National Park, Kentucky, USA. Hydrobiologia 520:19–28.

Gruberts, D., I. Druvietis, E. Parele, J. Paidere, A. Poppels, J. Prieditis, and A. Skute. 2007. Impact of hydrology on aquatic communities of floodplain lakes along the Daugava River (Latvia). Hydrobiologia 584:223–237.

Gücker, B. and I.G. Böechat. 2004. Stream morphology controls ammonium retention in tropical headwaters. Ecology 85:2818–2827.

Gupta, A. 2002. Large Rivers. Geomorphology 44:173–174.

Hack, J.T. 1957. Studies of longitudinal stream profiles in Virginia and Maryland. U.S. Geological Survey Professional Paper, 45–97.

Hall, R.O., Jr., J.L. Tank, and M.F. Dybdahl. 2003. Exotic snails dominate nitrogen and carbon cycling in a highly productive system. Frontiers in Ecology and the Environment 1:407–411.

Hamilton, S.K. and W.M. Lewis, Jr. 1987. Causes of seasonality in the chemistry of a lake on the Orinoco River floodplain Venezuela. Limnology and Oceanography 32:1277–1290.

Hamilton, S.K., W.M. Lewis, Jr., and S.J. Sippel. 1992. Energy sources for aquatic animals in the Orinoco River floodplain: Evidence from stable isotopes. Oecologia 89:324–330.

Hannah, D.M., P.J. Wood, and J.P. Sadler. 2004. Ecohydrology and hydroecology: A "new paradigm"? Hydrological Processes 18:3439–3445.

Hansmann, E.W. and H.K. Phinney. 1973. Effects of logging on periphyton in coastal streams of Oregon. Ecology 54:194–199.

Harrison, S., D.D. Murphy, and P.R. Ehrlich. 1988. Distribution of the bay checkerspot butterfly, *Euphydryas editha bayensis:* Evidence for a metapopulation model. American Naturalist 132:360–382.

Hart, D.D. 1983. The importance of competitive interactions within stream populations and communities. Pages 99–136 *in*: J.R. Barnes and G.W. Minshall (eds.), Stream Ecology. Plenum Press, New York.

Harte, J. 2002. Toward a Synthesis of the Newtonian and Darwinian Worldviews. Physics Today 55:29–43.

Harvey, C.J., B.J. Peterson, W.B. Bowden, A.E. Hershey, M. Miller, L.A. Deegan, and J.C. Finlay. 1998. Biological responses to fertilization of Oksrukuyik Creek, a tundra stream. Journal of the North American Benthological Society 17:190–209.

Havel, J.E., E.M. Eisenbacher, and A.A. Black. 2000. Diversity of crustacean zooplankton in riparian wetlands: Colonization and egg banks. Aquatic Ecology 34:63–76.

Hawkes, H.A. 1975. River zonation and classification. Pages 312–374 *in*: B.A. Whitton (ed.), River Ecology. Blackwell Science Publishers, Oxford, UK.

Hedges, J.I., W.A. Clark, P.D. Quay, J.E. Richey, A.H. Devol, and U. de M.Santos. 1986. Composition and fluxes of particulate organic material in the Amazon River. Limnology and Oceanography 31:717–738.

Hein, T., C. Baranyi, G.J. Herndl, W. Wanek, and F. Schiemer. 2003. Allochthonous and autochthonous particulate organic matter in floodplains of the River Danube: The importance of hydrological connectivity. Freshwater Biology 48:220–232.

Hein, T., W. Reckendorfer, J. Thorp, and F. Schiemer. 2005. The role of slackwater areas and the hydrologic exchange for biogeochemical processes in river corridors: Examples from the Austrian Danube. (Large Rivers Vol. 15), Archiv für Hydrobiologie (Suppl.) 155:425–442.

Hesse, L.W. and G.E. Mestl. 1993. An alternative approach for the Missouri River based on the precontrol condition. North American Journal of Fisheries Management 13:360–366.

Hetrick, N.J., M.A. Brusven, W.R. Meehan, and T.C. Bjornn. 1998. Changes in solar input, water temperature, periphyton abundance, and allochthonous input and storage after canopy removal along two small salmon streams in southeast Alaska. Transactions of the American Fisheries Society 127:859–875.

Hildrew, A.G. 1996. Whole river ecology. Spatial scale and heterogeneity in in the ecology of running waters. Archiv für Hydrobiologie 113:25–43.

Hildrew, A.G. and P.S. Giller. 1993. Patchiness, species interactions and disturbance in stream benthos. Pages 21–62 *in*: P.S. Giller, A.G. Hildrew, and D.G. Raffaelli (eds.), Aquatic ecology: Scale, pattern and process. Blackwell Scientific Publications.

Hill, M.O. 1979. TWINSPAN – a FORTRAN program for arranging mulitvariate data in an ordered two way table by classification of the individuals and the attributes. Cornell University, Department of Ecology and Systematics.

Høberg, P., M. Lindholm, L. Ramberg, and D.O. Hessen. 2002. Aquatic food web dynamics on a floodplain in the Okavango delta. Hydrobiologia 470:23–30.

Holland, M.M. 1988. SCOPE/MAB technical consultants on landscape boundaries: Report of SCOPE/MAB workshop on ecotones. Biology International, Special Issue 17:47–106.

Hooke, J.M. 1988. (Ed.) Geomorphology in environmental planning. Wiley, Chichester.

Horton, R.E. 1945. Erosional development of streams and their drainage basins: Hydrophysical approach to quantitative morphology. Geological Society of America Bulletin 56:275–370.

Huet, M. 1954. Biologie, profils en long et en travers des eaux courantes. Bulletin Francais de la Peche et de la Pisciculture 175:41–53.

Huff, D.R. 1986. Phytoplankton communities in navigation pool No. 7 of the Upper Mississippi River. Hydrobiology 136: 47–56.

Hughes, R.M., S.G. Paulsen, and J.L. Stoddard. 2000. EMAP-surface waters: A multi assemblage, probability survey of ecological integrity in the U.S.A. Hydrobiologia 422/423:430–443.

Hughes, V., M.C. Thoms, S. Nicol, and J. Koehn. 2008. Physical–ecological interactions in a lowland river system: Large wood, hydraulic complexity and native fish associations in the River Murray, Australia. Pages 387–404 *in*: P.J. Wood, D.M. Hannah, and J.P. Sadler (eds.), Hydrecology and ecohydrology: Past, present, and future. Wiley, Chichester, UK.

Humphries, P., R.A. Cook, A.J. Richardson, and L.G. Serafini. 2006. Creating a disturbance: Manipulating slackwaters in a lowland river. River Research and Applications 22:525–542.

Humphries, P., A.J. King, and J.D. Koehn. 1999. Fish, flows and flood plains: Links between freshwater fishes and their environment in the Murray-Darling River system, Australia. Environmental Biology of Fishes 56:129–151.

Humphries, P., G.S. Luciano, and A.J. King. 2002. River regulation and fish larvae: Variation through space and time. Freshwater Biology 47:1307–1331.

Hurley, K.L., R.J. Sheehan, R.C. Heidinger, P.S. Wills, and B. Clevenstine. 2004. Habitat use by middle Mississippi River pallid sturgeon. Transactions of the American Fisheries Society 133:1033–1041.

Hutchinson, G.E. 1958. Concluding remarks. Cold Springs Harbor Symposium in Quantitative Biology 22:415–427.

Hutchinson, G.E. 1961. The paradox of the plankton. American Naturalist 95:137–146.

Hynes, H.B.N. 1970. The ecology of running waters. University of Toronto Press, Toronto, Ont., Canada. 555 pp.

Hynes, H.B.N. 1983. Groundwater and stream ecology. Hydrobiologia 100:93–99.

Illies, J. 1961. Versuch einer allgemein biozönotischen Gliederung der Fliessgewässer. Internationale Revue der Gesamten Hydrobiologie 46:205–213.

Illies, J. and L. Botosaneanu. 1963. Problèmes et méthodes de la classification et de la zonation écologique des eaux courantes, considerées surtout du point de vue faunistique. Mitteilungen Internationale Vereinigung für theoretische und angewandte Limnologie 12, 57 pp.

Iwata, T., S. Nakano, and M. Inoue. 2003. Impacts of past riparian deforestation on stream communities in a tropical rain forest in Borneo. Ecological Applications 13:461–473.

Jack, J.D. and J.H. Thorp. 2000. Effects of the benthic suspension feeder *Dreissena polymorpha* on zooplankton in a large river. Freshwater Biology 44:569–579.

Jack, J.D. and J.H. Thorp. 2002. Impacts of fish predation on an Ohio River zooplankton community. Journal of Plankton Research 24:119–127.

Jackson, C.R., P.F. Churchill, and E.E. Roden. 2001. Successional changes in bacterial assemblage structure during epilithic biofilm development. Ecology 82:555–566.

Jackson, J.K., A.D. Huryn, D.L. Strayer, D.L. Courtemanch, and B.W. Sweeney. 2005. Atlantic Coast rivers of the northeastern United States. Pages 21–122 *in*: A.C. Benke and C.E. Cushing (eds.), Rivers of North America. Elsevier, San Diego, CA.

Jansson, R., U. Zinko, D.M. Merritt, and C. Christer. 2005. Hydrochory increases riparian plant species richness: A comparison between a free-flowing and a regulated river. Journal of Ecology 93:1094–1103.

Jaromilek, I., V. Banasova, H. Otahelova, and M. Zaliberova. 2001. Nine year succession of the herbaceous floodplain vegetation in the Morava River. Ekologia 20(suppl. 2):92–100.

Jenkins, K.M. and A.J. Boulton. 2003. Connectivity in a dryland river: Short-term macroinvertebrate recruitment following floodplain inundation. Ecology 84:2708–2723.

Jepsen, D.B. and K.O. Winemiller. 2002. Structure of tropical river food webs revealed by stable isotope ratios. Oikos 96:46–55.

Johnson, B.L., W.R. Richardson, and T.J. Naimo. 1995. Past, present, and future concepts in large river ecology. BioScience 45:134–141.

Joly, P. and A. Morand. 1994. Theoretical habitat templetes, species traits, and species richness – amphibians in the upper Rhone River and its floodplains. Freshwater Biology 31:455–468.

Jungwirth, M., S. Muhar, and S. Schmutz. 2000. Fundamentals of fish ecological integrity and their relation to the extended serial discontinuity concept. Hydrobiologia 422/423:85–97.

Junk, W.J. 1980. Areas inundáveis – Um desafío para Limnología. Acta Amazonica 10:775–795.

Junk, W.J., P.B. Bayley, and R.E. Sparks. 1989. The flood-pulse concept in river-floodplain systems. Pages 110–127 *in*: D.P. Dodge (ed.), Proceedings of the International Large River Symposium (LARS). Canadian Special Publication in Fishers and Aquatic Sciences, 106.

Junk, W.J. and K.M. Wantzen. 2004. The flood pulse concept: New aspects, approaches, and applications – an update. Pages 117–149 *in*: R.L. Welcomme and R.L. Petr (eds.), Proceedings of the second international symposium on the management of large rivers for fisheries, Volume 2. Food and Agriculture Organization & Mekong River Commission. FAO Regional Office for Asia and the Pacific, Bangkok. RAP Publication 2004/16.

Karaus, U., L. Alder, and K. Tockner. 2005. "Concave islands": Habitat heterogeneity of parafluvial ponds in a gravel-bed river. Wetlands 25:26–37.

Karrasch, B., M. Mehrens, Y. Eosenloecher, and K. Peters. 2001. The dynamics of phytoplankton, bacteria, and heterotrophic flagellates at two banks near Magdeburg in the river Elbe (Germany). Limnologica 31:93–107.

Karrenberg, S., P.J. Edwards, and J. Kollmann. 2002. The life history of Salicaceae living in the active zone of floodplains. Freshwater Biology 47:733–748.

Kay, J.J., H.A. Regier, M. Boyles, and G. Francis. 1999. An ecosystem approach for sustainability: Addressing the challenge for complexity. Futures 31:721–742.

Kellerhals, R., M. Church, and D.I. Bray. 1976. Classification and analysis of river processes. Journal of the Hydraulics Division HY 7:813–829.

Kemp, M.J. and W.K. Dodds. 2002. The influence of ammonium, nitrate, and dissolved oxygen concentrations on uptake, nitrification, and denitrification rates associated with prairie stream substrata. Limnology and Oceanography 47:1380–1393.

Kendall, C., S.R. Silva, and V.J. Kelly. 2001. Carbon and nitrogen isotopic composition of particulate organic matter in four large river systems across the United States. Hydrological Processes 15:1301–1346.

Kessler, R.K. 1998. The functional importance of shoreline habitats to fishes of the Ohio River. Ph.D. dissertation, University of Louisville, Louisville, KY.

Kiffney, PM, J.S. Richardson, and J.P. Bull. 2003. Responses of periphyton and insects to experimental manipulation of riparian buffer width along forest streams. Journal of Applied Ecology 40:1060–1076.

Kiffney, P.M., J.S. Richardson, and J.P. Bull. 2004. Establishing light as a causal mechanism structuring stream communities in response to experimental manipulation of riparian buffer width. Journal of the North American Benthological Society 23:542–555.

King, A.W., 1997. Hierarchy theory: A guide to system structure for wildlife biologists. Pages 185–212 *in*: Bissonette J.A. (ed.), Wildlife and Landscape Ecology: Effects of Pattern and Scale. Springer-Verlag, New York.

King, J.M. and M.D. Louw. 1998. Instream flow assessments for regulated rivers in South Africa using the Building Block Methodology. Aquatic Ecosystem Health and Management 1:109–124.

Knight, G.L. and K.B. Gido. 2005. Habitat use and susceptibility to predation of four prairie stream fishes: Implications for conservation of the endangered Topeka shiner. Copeia 2005:38–47.

Knighton, D., 1984. Fluvial Forms and Processes. Edward Arnold, London.

Knowlton, M.F. and J.R. Jones. 1997. Trophic status of Missouri River floodplain lakes in relation to basin type and connectivity. Wetlands 17:468–475.

Knuepfer, P.L.K. and J.K. Petersen. 2002. Geomorphology in the public eye: Policy issues, education and the public. Geomorphology 47:95–105.

Koel, T.M. 2004. Spatial variation in fish species richness of the Upper Mississippi River system. Transactions of the American Fisheries Society 133:984–1003.

Kohyama, T. 2005. Scaling up from shifting-gap mosaic to geographic distribution in the modeling of forest dynamics. Ecological Research 20:305–312.

Kondolf, G.M. and P.W. Downs. 1996. Catchment approach to planning channel restoration Pages 129–147 *in*: A. Brookes and F.D. Shields (eds.), River Channel Restoration: Guiding Principles for Sustainable Projects. Wiley, Chichester, UK.

Kondolf, G.M., D.R. Montgomery, H. Piegay, and L. Schmitt. 2003. Geomorphic classification of rivers and streams. Pages 171–204 *in*: G.M. Kondolf and H. Piegay (eds.), Tools in Fluvial Geomorphology. Wiley, Chichester, UK.

Koschorreck, M. and A. Darwich. 2003. Nitrogen dynamics in seasonally flooded soils in the Amazon floodplain. Wetlands Ecology and Management 11:317–330.

Kotliar, N.B. and J.A. Wiens. 1990. Multiple scales of patchiness and patch structure: A hierarchical framework for the study of heterogeneity. Oikos 59:253–260.

Krause, A.E., K.A. Frank, D.M. Mason, R.E. Ulanowicz, and W.W. Taylor. 2003. Compartments revealed in food-web structure. Nature 426:282–285.

Kuehne, R.A. 1962. A classification of streams, illustrated by fish distribution in an eastern Kentucky creek. Ecology 43:608–614.

Lake, P.S. 2000. Disturbance, patchiness, and diversity in streams. Journal of the North American Benthological Society 19:573–592.

Lake, P.S. 2003. Ecological effects of perturbation by drought in flowing waters. Freshwater Biology 48:1161–1172.

Lamberti, G.A., S.V. Gregory, C.P. Hawkins, R.C. Wildman, L.R. Ashkenas, and D.M. Denicola. 1992. Plant herbivore interactions in streams near Mount St. Helens. Freshwater Biology 27:237–247.

Lancaster, J. 2000. Geometric scaling of microhabitat patches and their efficacy as refugia during disturbance. Journal of Animal Ecology 69:442–457.

Lancaster, J. and L.R. Belyea. 1997. Nested hierarchies and scale-dependence of mechanisms of flow refugium use. Journal of the North American Benthological Society 16:221–238.

Lane, E.W. 1955. The importance of fluvial morphology in hydraulic engineering. Proceedings of the American Society of Civil Engineering 81, 745:1–17.

Lane, S.N. and K.S. Richards. 1997. Linking river channel form and process: Time, space and causality revisited. Earth Surface Processes and Landforms, 22:249–260.

Lanka, R.P., W.A. Hubert, and T.A. Wesche. 1987. Relations of geomorphology to stream habitat and trout standing stock in small Rocky Mountain streams. Transactions of the American Fisheries Society 116:21–28.

Latterell, J.J., J.S. Bechtold, T.C. O'Keefe, R. Van Pelt, and R.J. Naiman. 2006. Dynamic patch mosaics and channel movement in an unconfined river valley of the Olympic Mountains. Freshwater Biology 51:523–544.

Layman, C.A., K.O. Winemiller, and D.A. Arrington. 2005. Describing the structure and function of a Neotropical river food web using stable isotopes, stomach contents, and functional experiments. Pages 395–406 *in*: P.C. de Ruiter, V. Wolters, and J.C. Moore (eds.), Dynamic Food Webs: Multispecies Assemblages, Ecosystem Development and Environmental Change, Elsevier, Amsterdam.

LeFebvre, S., P. Marmonier, and G. Pinay. 2004. Stream regulation and nitrogen dynamics in sediment interstices: Comparison of natural and straightened sectors of a third-order stream. River Research and Applications 20:499–512.

Legendre, P. 1993. Spatial autocorrelation: Trouble or new paradigm? Ecology 74:1659–1673.

Lehtinen, R.M., N.D. Mundahl, and J.C. Madejczyk. 1997. Autumn use of woody snags by fishes in backwater and channel border habitats of a large river. Environmental Biology of Fishes 49:7–19.

Leigh, C. and F. Sheldon. 2008. Hydrological changes and ecological impacts associated with water resource development in large floodplain rivers in the Australian tropics. River Research and Applications.

Leland, H.V. and S.D. Porter. 2000. Distribution of benthic algae in the upper Illinois River basin in relation to geology and land use. Freshwater Biology 44:279–301.

Leopold, L.B. and M.G. Wolman. 1957. River channel patterns: Braided, meandering and straight. United States Geological Survey Professional Paper, 282B: 51.

Leopold, L.D., M.G. Wolman, and J.P. Miller. 1964. Fluvial Processes in Geomorphology. W.H. Freeman & Company, New York.

Lepori, F., D. Palm, E. Braenaes, and B. Malmqvist. 2005. Does restoration of structural heterogeneity in streams enhance fish and macroinvertebrate diversity? Ecological Applications 15:2060–2071.

Levin, S.A. 1992. The problem of pattern and scale in ecology. Ecology 73:1943–1967.

Levins, R. 1969. Some demographic and genetic consequences of environmental heterogeneity for biological control. Bulletin of the Entomological Society of America 15:237–240.

Lewis, W.M. 1988. Primary production in the Orinoco River. Ecology 69:679–692.

Lewis, W.M. Jr., S.K. Hamilton, M.A. Lasi, M. Rodríguez, and J.F. Saunders, III. 2000. Ecological determinism on the Orinoco floodplain. BioScience 50:681–692.

Lewis, W.M. Jr., S.K. Hamiliton, M.A. Rodríguez, J.F. Saunders, III, and M.A. Lasi. 2001. Foodweb analysis of the Orinoco floodplain based on production estimates and stable isotope data. Journal of the North American Benthological Society 20:241–254.

Lewis, W.M. Jr., F.H. Weibezahn, J.F. Saunders, III, and S.K. Hamilton. 1990. The Orinoco River as an ecological system. Interciencia 15:346–357.

Lin, C, W. Chou, and W. Lin. 2002. Modeling the width and placement of riparian vegetated buffer strips: A case study on the Chi-Jia-Wang Stream, Taiwan. Journal of Environmental Management 66:269–280.

Lindeman, R.E., 1942. Trophic-dynamic aspect of ecology. Ecology 23:399–418.

Lindholm, M. and D.O. Hessen. 2007. Zooplankton succession on seasonal floodplains: Surfing on a wave of food. Hydrobiologia 592:95–104.

Logue, J.B., C.T. Robinson, C. Meier, and J.R. Van der Meer. 2004. Relationship between organic matter, bacteria composition, and the ecosystem metabolism of alpine streams. Limnology and Ocenaography 49:2001–2010.

Lotspeich, F.B., 1980. Watersheds as the basic ecosystem: This conceptual framework provides a basis for a natural classification system. Water Resources Bulletin 16:581–586.

Lorang, M.S., D.C. Whited, F.R. Hauer, J.S. Kimball, and J.A. Stanford. 2005. Using airborne multispectral imagery to evaluate geomorphic work across floodplains of gravel-bed rivers. Ecological Applications 15:1209–1222.

Lorenz, C.M., G.M. Van Dijk, A.G.M. Van Hattum, and W.P. Cofino. 1997. Concepts in river ecology: Implications for indicator development. Regulated Rivers: Research and Management 13:501–516.

Low, A.F. and J.D. Lyons. 1995. Effects of flow regulation and restriction of passage due to hydroelectric project operation on the structure of fish and invertebrate communities in Wisonsin's large river systems. Wisconsin Department of Natural Resources preliminary report, Madison, WI.

Ludwig, J.A. and J.M. Cornelis. 1987. Locating discontinuities along ecological gradients. Ecology 68:448–449.

Lyautey, E., C.R. Jackson, J. Cayrou, J.L. Rols, and F. Garabetian. 2005. Bacterial community succession in natural river biofilm assemblages. Microbial Ecology 50:589–601.

Lytle, D.A. 2002. Flash floods and aquatic insect life-history evolution: Evaluation of multiple models. Ecology 83:370–385.

Maddock, I., M.C. Thoms, K. Jonson, F. Dyer, and M. Lintermanns. 2004. Identifying the influence of channel morphology on physical habitat svailability for native fish: Application to the two-spined blackfish (Gadopsis bispinosus) in the Cotter River, Australia. Marine and Freshwater Research 55:173–184.

Magurran, A. 2004. Measuring Biological Diversity. Blackwell Scientific, USA.

Maine, M.A., N.L. Sune, and C. Bonetto. 2004. Nutrient concentrations in the Middle Parana River: Effect of the floodplain lakes. Archiv für Hydrobiologie 160:85–103.

Malard, F., M. Alain, U. Uehlinger, and J.V. Ward. 2001. Thermal heterogeneity in the hyporheic zone of a glacial floodplain. Canadian Journal of Fisheries and Aquatic Sciences 58:1319–1335.

Malard, F., K. Tockner, and J.V. Ward. 1999. Shifting dominance of subcatchment water sources and flow paths in a glacial floodplain, Val Roseg, Switzerland. Arctic, Antarctic, and Alpine Research 31:135–150.

Malard, F., U. Uehlinger, R. Zah, and K. Tockner. 2006. Flood-pulse and riverscape dynamics in a braided glacial river. Ecology 87:704–716.

Malcolm, I.A., C. Soulsby, A.F. Youngson, and J. Petry. 2003. Heterogeneity in ground water-surface water interactions in the hyporheic zone of a salmonid spawning stream. Hydrological Processes 17:601–617.

Malmqvist, B. 2002. Aquatic invertebrates in riverine landscapes. Freshwater Biology 47:679–694.

Mandelbrot, B. 1982. The Fractal Geometry of Nature. W.H. Freeman Co., New York, NY.

Mander, U., A. Kull, V. Kuusemets, and T. Tamm. 2000. Nutrient runoff dynamics in a rural catchment: Influence of land-use changes, climatic fluctuations and ecotechnological measures. Ecological Engineering 14:405–417.

Margules, C.R. and R. Pressey. 2000. Systematic conservation planning. Nature 405:243–253.

Marks, J.C., M.E. Power, and M.S. Parker. 2000. Flood disturbance, algal productivity, and interannual variation in food chain length. Oikos 90:20–27.

Marshall, J.C., F. Sheldon, M. Thoms, and S. Choy. 2006. The macroinvertebrate fauna of an Australian dryland river: Spatial and temporal patterns and environmental relationships. Marine and Freshwater Research 57:61–74.

Mayr, E. 1970. Populations, Species, and Evolution. Harvard University Press, Cambridge, Massachusetts. 453 p.

Mermillod-Blondin, F., M. Gerino, S. Sauvage, and M.C. de Chatelliers. 2004. Influence of non-trophic interactions between benthic invertebrates on river sediment processes: A microcosm study. Canadian Journal of Fisheries and Aquatic Sciences 61:1817–1831.

Meyer, J.L. 1996. Conserving ecosystem function. Pages 136–145 *in*: T.A. Pickett, R.S. Ostfield, M. Shachak, and G.E. Likens (eds.), The Ecological Basis of Conservation: Heterogeneity, Ecosystems, and Biodiversity. Chapman and Hall, New York, USA.

Mihuc, T.B. 1997. The functional trophic role of lotic primary consumers: Generalist versus specialist strategies. Freshwater Biology 37:455–462.

Mihuc, T.B. and G.W. Minshall. 2005. The trophic basis of reference and post-fire stream food webs 10 years after wildfire in Yellowstone National Park. Aquatic Sciences 67:541–548.

Miller, A.J. 1995. Valley morphology and boundary conditions influencing spatial patterns of flood flow. Pages 57–82 *in*: J.E. Costa, A.J. Miller, K.W. Potter, and P.R. Wilcock (eds.), Natural and Anthropogenic Influences in Fluvial Geomorphology. American Geophysical Union, Washington.

Minshall, G.W., J.T. Brock, and T.W. LaPoint. 1982. Characterization and dynamics of benthic organic matter and invertebrate functional feeding group relationships in the upper Salmon River, Idaho (USA). Internationale Revue der Gesamten Hydrobiologie 67:793–820.

Minshall, G.W., K.W. Cummins, R.C. Petersen, C.E. Cushing, D.A. Bruns, J.R. Sedell, and R.L. Vannote. 1985. Developments in stream ecosystem theory. Canadian Journal of Fisheries and Aquatic Sciences 42:1045–1055.

Minshall, G.W., R.C. Petersen, K.W. Cummins, T.L. Bott, J.R. Sedell, C.E. Cushing, and R.L. Vannote. 1983. Interbiome comparison of stream ecosystem dynamics. Ecological Monographs 53:1–25.

Miranda, L.E. 2005. Fish assemblages in oxbow lakes relative to connectivity with the Mississippi River. Transactions of the American Fisheries Society 134:1480–1489.

Molles, M.C., Jr., C.S. Clifford, L.M. Ellis, H.M. Valett, and C.N. Dahm. 1998. Managing flooding for riparian ecosystem restoration. Bioscience 48:749–756.

Montgomery, D.R. 1999. Process domains and the river continuum concept. Journal of the American Water Resources Association 35:397–410.

Montgomery, D.R. and J.M. Buffington. 1997. Channel-reach morphology in mountain drainage basins. Geological Society of America Bulletin 109:596–611.

Montgomery, D.R. and J.M. Buffington. 1998. Channel processes, classification, and response. Pages 13–42 *in*: R.J. Naiman and R.E. Bilby (eds.), River Ecology and Management: Lessons from the Pacific Coastal Ecoregion, Springer, Verlag.

Moon, B.P., A.W. van Niekerk, G.L. Heritage, K.H. Rogers, and C.S. James. 1997. A geomorphological approach to the management of rivers in the Kruger National Park: The case of the Sabie River. Transactions of the Institute of British Geographers 22:31–48.

Moore, S.L. and J.H. Thorp. 2008. Environmental constraints on larval and juvenile fishes in a hydrogeomorphically dynamic prairie river. River Research and Applications 24:237–248.

Morisawa, M. 1968. Streams: Their Dynamics and Morphology. McGraw Hill Book Company, New York.

Morris, C. (ed.). 1992. Academic Press Dictionary of Science and Technolgy. Academic Press (Elsevier), San Diego.

Mosley, M.P. 1981. The classification and characterisation of rivers. Pages 295–320 *in*: K.S. Richards (ed.), Rivers: Environment and Process. Blackwell, Oxford.

Muirhead, J.R. and H.J. MacIssac. 2005. Development of inland lakes as hubs in an invasion network. Journal of Applied Ecology 42:80–90.

Mulholland, P.J., J.L. Tank, J.R. Webster, W.B. Bowden, W.K. Dodds, S.V. Gregory, N.B. Grimm, S.K. Hamilton, S.L. Johnson, E. Marti, W.H. McDowell, J.L. Merriam, J.L. Meyer, B.J. Peterson, H.M. Valett, and

W.M. Wollheim. 2002. Can uptake length in streams be determined by nutrient addition experiments? Results from an interbiome comparison study. Journal of the North American Benthological Society 21:544–560.

Nakano, S., H. Miyasaka, and N. Kuhara. 1999. Terrestrial-aquatic linkages: Riparian arthropod inputs alter trophic cascades in a stream food web. Ecology 80:2435–2441.

Naiman, R.J. 1997. Large animals and system-level characteristics in river corridors. BioScience 47:521–529.

Naiman, R.J., H. Decamps, and M.E. McClain. 2005. Riparia: Ecology, Conservation, and Management of Streamside Communities. Elsevier, San Diego, USA 430 p.

Naiman, R.J., D.G. Lonzarich, T.J. Beechie, and S.C. Ralph. 1992. General principles of classification and the assessment of conservation in rivers. Pages 93–123 *in*: P.J. Boon, P. Calow, and G.E. Petts (eds.), River Conservation and Management. American Geophysical Union, Washington D.C.

Naiman, R.J., J.M. Melillo, and J.E. Hobbie. 1986. Ecosystem alteration for boreal forest streams by beaver (*Castor canadensis*). Ecology 67:1254–1269.

Nanson, G.C. and A.D. Knighton. 1996. Anabranching rivers: Their cause, character and classification. Earth Surface Processes and Landforms 21:217–239.

Newbold, J.D. 1992. Cycles and spirals of nutrients. Pages 379–408 *in*: P. Calow and G.E. Petts (eds.), The Rivers Handbook, Volume. 1. Blackwell Science, Oxford, UK.

Newbold, J.D., R.V. O'Neill, J.W. Elwood, and W. Van Winkle. 1982. Nutrient spiraling in streams: Implications for nutrient and invertebrate activity. American Naturalist 120:628–652.

Newman, M.E.J. 2003. The structure and function of complex networks. SIAM Review 45:167–256.

Newson, M.D., M.J. Clark, D.A. Sear, and A. Brookes. 1998. The geomorphological basis for classifying rivers. Aquatic Conservation: Marine and Freshwater Ecosystems 8:415–430.

Nichols, S.J., D.W. Schloesser, and P.L. Hudson. 1989. Submersed macrophytes communities before and after an episodic ice jam in the St. Clair and Detroit rivers. Canadian Journal of Botany 67:2364–2370.

Nielsen, L.P., A. Enrich-Prast, and F.A. Esteves. 2004. Pathways of organic matter mineralization and nitrogen regeneration in the sediment of five tropical lakes. Acta Limnologica Brasiliensia 16:193–202.

Nilsson, C., M. Gardfjell, and C. Grelsson. 1991. Importance of hydrochory in structuring plant communities along rivers. Canadian Journal of Botany 69:2631–2633.

Norris, R.H. and A. Georges. 1993. Analysis and interpretation of benthic surveys. Pages 234–286 *in*: D.M. Rosenberg and V.H. Resh eds. Freshwater Biomonitoring and Benthic Macroinvertebrates. Chapman and Hall, New York.

Norris, R.H., S. Linke, I. Prosser, W.J. Young, P. Liston, N. Bauer, N. Sloane, F. Dyer, and M.C. Thoms. 2007. Very-broadscale assessment of human impacts on river condition. Freshwater Biology 52:959–976.

Norris, R.H. and C.P. Hawkins. 2000. Monitoring river health. Hydrobiologia 435:5–17.

Norris, R.H. and K.H. Norris. 1995. The need for a biological assessment of water quality: Australian perspective. Australian Journal of Ecology 20:1–6.

O'Donnell, J.A. and J.B. Jones, Jr. 2006. Nitrogen retention in the riparian zone of catchments underlain by discontinuous permafrost. Freshwater Biology 51:854–864.

Olsen, A.R., J. Sedransk, D. Edward, C.A. Gotway, W. Ligget, S. Rathbun, K.H. Reckhow, and L.J. Young. 1999. Statistical issues for monitoring ecological and natural resources in the United States. Environmental Monitoring and Assessment 54:1–45.

O'Neill, R.V. 1989. Perspectives in hierarchy and scale. Pages 140–156 *in*: J. Roughgarden, R.M. May, and S.A. Levin (eds.), Perspectives in Ecological Theory. Princeton University Press, New Jersey.

O'Neill, R.V., and A.W. King. 1998. Homage to St. Michael; or, why are there so many books on scale? Pages 3–15 *in*: D.L. Peterson and V.T. Parker (eds.), Ecological Scale: Theory and Applications. Columbia University Press, New York.

O'Neill, R.V., D.L. DeAngelis, J.B. Waide, and T.F.H. Allen. 1986. A Hierarchical Concept of Ecosystems, Princeton University Press, Princeton, New Jersey.

O'Neill, R.V., R.H. Gardner, B.T. Milne, M.G. Turner, and B. Jackson. 1991. Heterogeneity and spatial hierarchies. Pages 85–96 *in*: J. Kolasa and S.T.A. Pickett (eds.), Ecological Heterogeneity. Springer-Verlag, New York.

O'Neill, R.V., A.R. Johnson, and A.W. King. 1989. A hierarchical framework for the analysis of scale. Landscape Ecology 3:193–205.

Opsahl, S. and R. Benner. 1998. Photochemical reactivity of dissolved lignin in river and ocean waters. Limnology and Oceanography 43:1297–1304.

Orr, C.H., K.L. Rogers, and E.H. Stanley. 2006. Channel morphology and P uptake following removal of a small dam. Journal of the North American Benthological Society 25:556–568.

Osmundson, D.B., R.J. Ryel, V.L. Lamarra, and J. Pitlick. 2002. Flow-sediment-biota relations: Implications for river regulation effects on native fish abundance. Ecological Applications 12:1719–1739.

Overton, I.C., P.G. Slavich, M.M. Lewis, and G.R. Walker. 2006. Modelling vegetation health from the interaction of saline groundwater and flooding on the Chowilla floodplain, South Australia. Australian Journal of Botany 54:207–220.

Pace, M.L., J.J. Cole, S.R. Carpenter, and J.F. Kitchell. 1999. Trophic cascades revealed in diverse ecosystems. Trends in Ecology and Evolution 14:483–487.

Pace, M.L., S.E.G. Findlay, and D. Fischer. 1998. Effects of an invasive bivalve on the zooplankton community of the Hudson River. Freshwater Biology 39:103–116.

Pace, M.L., S.E.G. Findlay, and D. Lints. 1992. Zooplankton in advective environments: The Hudson River community and a comparative analysis. Canadian Journal of Fisheries and Aquatic Science 49:1060–1069.

Palmer, L. 1976. River management criteria for Oregon and Washington. Pages 141–153 *in*: D.R. Coates (ed.), Geomorphology and engineering. Halsted Press, Stroudsburg, Pennsylvania.

Palmer, M.A. and E.S. Bernhardt. 2006. Hydroecology and river restoration: Ripe for research and synthesis. Water Resources and Research 42: W03S07.

Palmer, M.A., C.M. Swan, K. Nelson, P. Silver, and R. Alvestad. 2000. Streambed landscapes: Evidence that stream invertebrates respond to the type and spatial arrangement of patches. Landscape Ecology 15:563–576.

Paine, R.T. 1966. Food web complexity and species diversity. American Naturalist 100:65–75.

Paine, R.T. and S.A. Levin. 1981. Intertidal landscapes: Disturbance and the dynamics of pattern. Ecological Monographs 51:145–178.

Parsons, M.E. and M.C. Thoms. 2007. Hierarchical patterns of physical-biological associations in river ecosystems. Geomorphology 89:127–146.

Parsons, M., M.C. Thoms, and R.H. Norris. 2003. Scales of macroinvertebrate distribution in relation to the hierarchical organization of river systems. Journal of the North American Benthological Society 22:106–122.

Parsons, M.E., M.C. Thoms, and R.H. Norris. 2004a. Using hierarchy to select scales of measurement in multiscale studies of stream macroinvertebrates. Journal of the North American Benthological Society 23:157–170.

Parsons, M.E., M.C. Thoms, and R.H. Norris. 2004b. Development of a standardised approach to river habitat assessment in Australia. Environmental Assessment and Monitoring 98:109–130.

Partridge, T.C. and R.R. Maud. 1997. Geomorphic evolution of southern Africa since the Mesozoic. South African Journal of Geology 90:179–208.

Patil, G.P 2002. Multiscale advanced raster map analysis system: Network-based analysis of biological integrity in freshwater streams. Center for Statistical Ecology and Environmental Statistics, The Pennsylvania State University, University Park, Pennsylvania, USA.

Paustian, S.J., D.A. Marion, and D.F. Kelliher. 1984. Stream channel classification using large scale aerial photography for southeast Alaska watershed management. Renewable Resources Management Symposium Pages 670–677 American Society of Photogrammetry, Falls Church, VA.

Peckarsky, B.L. 1979. Biological interactions as determinants of distributions of benthic invertebrates within stony streams. Limnology and Oceanography 24:59–68.

Peckarsky, B.L., B.W. Taylor, and C.C. Caudill. 2000. Hydrologic and behavioural constraints on oviposition of stream insects: Implications for adult dispersal. Oecologia 125:186–200.

Pegg, M.A. and C.L. Pierce. 2002. Classification of reaches in the Missouri and lower Yellowstone Rivers based on flow characteristics. River Research and Applications 18:31–42.

Perry, J.A. and D.J. Schaeffer. 1987. The longitudinal distribution of riverine benthos: A river dis-continuum? Hydrobiologia 148:257–268.

Peterson, D.L. and V.T. Parker. 1998. Ecological Scale – Theory and Applications. Complexity in Ecological Systems New York, Columbia University Press.

Petry, P., P.B. Bayley, and D.F. Markle. 2003. Relationships between fish assemblages, macrophytes and environmental gradients in the Amazon River floodplain. Journal of Fish Biology 63:547–579.

Petts, G.E., 1984. Impounded Rivers. Wiley, Chichester, UK.

Petts, G.E. and C. Amoros. 1996. Fluvial hydrosystems. Chapman & Hall, London.

Pflieger, W.L. 1975. The fishes of Missouri. Missouri Department of Conservation, Columbia, South Carolina, USA.

Phillips, J.D., 1995. Biogeomorphology and landscape evolution: The problem of scale. Geomorphology 13:337–347.

Phillips, J.D. 1999. Earth Surface Systems: Complexity, Order, and Scale. Blackwell, Oxford.

Phillips, J.D. 2006. The perfect landscape. Geomorphology 84:159–169.

Pickett, S.T.A., M.L. Cadenasso, and T.L. Benning. 2003. Biotic and abiotic variability as key determinants of savanna heterogeneity at multiple spatiotemporal scales. Pages 22–40 *in*: J.T. Du Toit, K.H. Rogers, and H.C. Biggs, (eds.), The Rivers of Kruger National Park. South African Water Research Commission, Pretoria.

Pickett, S.T.A., W.R. Burch, and J.M. Grove. 1999. Interdisciplinary research: Maintaining the constructive impulse in a culture of criticism. Ecosystems 2:302–307.

Pickett, S.T.A. and P.S. White. (eds.). 1985. The Ecology of Natural Disturbance and Patch Dynamics. Academic Press, New York. 472 p.

Piegay, H., G. Bornette, A. Citterio, E. Herouin, B. Moulin, and C. Stratiotis. 2000. Channel instability as a control factor of silting dynamics and vegetation pattern within perifluvial aquatic zones. Hydrological Processes 14:3011–3029.

Pinay, G., C. Ruffinoni, S. Wondzell, and F. Gazelle. 1998. Change in groundwater nitrate concentration in a large river floodplain: Denitrification, uptake, or mixing? Journal of the North American Benthological Society 17:179–189.

Piscart, C., A. Lecerf, P. Usseglio-Polatera, J.C. Moreteau, and J.N. Beisel. 2005. Biodiversity patterns along a salinity gradient: The case of net-spinning caddisflies. Biodiversity and Conservation 14:2235–2249.

Poff, N.L. 1997. Landscape filters and species traits: Toward mechanistic understanding and prediction in stream ecology. Journal of the North American Benthological Society 16:391–409.

Poff, N.L., J.D. Allan, M.B. Bain, J.R. Karr, K.L. Prestegaard, B.D. Richter, R.E. Sparks, and J.C. Stromberg. 1997. The natural flow regime: A paradigm for river conservation and restoration. BioScience 47:769–784.

Poff, N.L., J.D. Olden, D.M. Pepin, and B.P. Bledsoe. 2006. Placing global stream flow variability in geographic and geomorphic contexts. River Research and Applications 22:149–166.

Poole, G.C. 2002. Fluvial landscape ecology: Addressing uniqueness within the river discontinuum. Freshwater Biology 47:641–660.

Poole, G.C., J.A. Stanford, S.W. Running, and C.A. Frissell. 2006. Multiscale geomorphic drivers of groundwater flow paths: Subsurface hydrologic dynamics and hyporheic habitat diversity. Journal of the North American Benthological Society 25(2):288–303.

Post, D.M. 2002a. The long and short of food-chain length. Trends in Ecology and Evolutions 17:269–277.

Post, D.M. 2002b. Using stable isotopes to estimate trophic position: Models, methods, and assumptions. Ecology 83:703–718.

Post, D.M., M.L. Pace, and N.G. Hairston. 2000. Ecosystem size determines food-chain length in lakes. Nature 405:1047–1049.

Post, D.M. and G. Takimoto. 2007. Proximate structural mechanisms for variation in food-chain length. Oikos 116:775–782.

Pouilly, M., T. Yunoki, C. Rosales, and L. Torres. 2004. Trophic structure of fish assemblages from Mamoré River floodplain lakes (Bolivia). Ecology of Freshwater Fish 13:245–257.

Power, M.E., A. Sun, M. Parker, W.E. Dietrich, and J.T. Wootton. 1995. Hydraulic food-chain models: An approach to the study of food-web dynamics in large rivers. BioScience 45:159–167.

Pringle, C.M., M.C. Freeman, and B.J. Freeman. 2000. Regional effects of hydrologic alterations on riverine macrobiota in the New World: Tropical-temperate comparisons. Bioscience 50:807–823.

Pringle, C.M., R.J. Naiman, G. Bretschko, J.R. Karr, M.W. Osgood, J.R. Webster, R.L. Welcomme, and M.J. Winterbourn. 1988. Patch dynamics in lotic systems: The stream as a mosaic. Journal of the North American Benthological Society 7:503–524.

Quinn, G.P. and M. Keough. 2002. Experimental design and data analysis for biologists. Cambridge University Press.

Quinn, J.M., R.B. Williamson, R.K. Smith, and M.L. Vickers. 1992. Effects of riparian grazing and channelisation on streams in Southland, New Zealand. 2. Benthic invertebrates. New Zealand Journal of Marine and Freshwater Research 26:259–273.

Rabalais, N.N., R.E. Turner, and W.J. Wiseman. 2002. Gulf of Mexico hypoxia, aka "the dead zone". Annual Review of Ecology and Systematics 33:235–263.

Raibley, P.T., K.S. Irons, K.D. Blodgett, and R.E. Sparks. 1997. Winter habitats used by largemouth bass in the Illinois River, a large river-floodplain ecosystem. North American Journal of Fisheries Management 17:401–412.

Rapport, A. 1985. Thinking about home environments: A conceptual framework. Pages 255–286 *in*: I. Altman and C.M. Werner (eds.), Home Environments. Plenum Press, New York.

Raven, P.J., N.T.H. Holmes, F.H. Dawson, and M. Everard. 1998a. Quality assessment using river habitat survey data. Aquatic Conservation: Marine and Freshwater Ecosystems 8:477–499.

Raven, P. J., N.T.H. Holmes, F.H. Dawson, P.J.A. Fox, M. Everard, I.R. Fozzard, and K.L. Rouen. 1998b. River habitat quality: The physical character of rivers and streams in the U.K. and Isle of Man. River Habitat Survey Report No., 2. Environment Agency, Bristol, England. 86 p.

Rayburg, S., M. Neave, M.C. Thoms, R. Nanson, E. Lenon, and T. Breen. 2006. A review of geomorphic classification schemes for rivers and floodplains. Technical Report to the Murray Darling Basin Commission, 40p.

Rees, G.N., G.O. Watson, D.S. Baldwin, and A.M. Mitchell. 2006. Variability in sediment microbial communities in a semi-permanent stream: Impact of drought. Journal of the North American Benthological Society 25:370–378.

Reid, M.A. and R.W. Ogden. 2006. Trend, variability or extreme event? The importance of long-term perspectives in river ecology. River Research and Applications 22:167–178.

Rejas, D., S. Declerck, J. Auwerkerken, P. Tak, and L. DeMeester. 2005. Plankton dynamics in a tropical floodplain lake: Fish, nutrients, and the relative importance of bottom-up and top-down control. Freshwater Biology 50:52–69.

Resh, V.H., A.V. Brown, A.P. Covich, M.E. Gurtz, H.W. Li, G.W. Minshall, S.R. Reice, A.L. Sheldon, J.B. Wallace, and R.C. Wissmar. 1988. The role of disturbance in stream ecology. Journal of the North American Benthological Society 7:433–455.

Reynolds, C.S. and J.-P. Descy. 1996. The production, biomass and structure of phytoplankton in large rivers. Archiv für Hydrobiologie, Supplement 113, Large Rivers 10:161–187.

Reynoldson, T.B., R.H. Norris, V.H. Resh, K.E. Day, and D.M. Rosenberg. 1997. The reference condition: A comparison of multimetric and multivariate approaches to assess water-quality impairment using benthic macroinvertebrates. Journal of the North American Benthological Society 16:833–852.

Rice, S.P., M.T. Greenwood, and C.B. Joyce. 2001. Tributaries, sediment sources, and the longitudinal organization of macroinvertebrate fauna along river systems. Canadian Journal of Fisheries and Aquatic Sciences 58:824–840.

Richards, K.R. 1982. Rivers: Form and process in alluvial channels. Cambridge University Press, Cambridge.

Richardson, W.B., E.A. Strauss, L.A. Bartsch, E.M. Monroe, J.C. Cavanaugh, L. Vingum, and D.M. Soballe. 2004. Denitrification in the Upper Mississippi River: Rates, controls and contribution to nitrate flux. Canadian Journal of Fisheries and Aquatic Sciences 61:1102–1112.

Richter, B.D., J.V. Baumgartner, J. Powell, and D.P. Braun. 1996. A method for assessing hydrologic alteration within ecosystems. Conservation Biology 10:1163–1174.

Rinne, J.N. 2005. Changes in fish assemblages, Verde River, Arizona, 1974–2003. Pages 115–136 *in*: J.N. Rinne, R.M. Hughes, and B. Calamusso (eds.), Historical Changes in Large River Fish Assemblages of the Americas, American Fisheries Society Symposium 45, Bethesda, Maryland, USA.

Rinne, J.N., J.R. Simms, and H. Blasius. 2005. Changes in hydrology and fish fauna in the Gila River, Arizona-New Mexico: Epitaph for native fish fauna? Pages 127–137 *in*: J.N. Rinne, R.M. Hughes, and B. Calamusso (eds.), Historical changes in large river fish assemblages of the Americas, American Fisheries Society Symposium 45, Bethesda, Maryland, USA.

Robertson, A.I., P. Bacon, and G. Heagney. 2001. The responses of floodplain primary production to flood frequency and timing. Journal of Applied Ecology 38:126–136.

Robertson, A.L. and A.M. Milner. 1999. Meiobenthic arthropod communities in new streams in Glacier Bay National Park, Alaska. Hydrobiologia 397:197–209.

Robertson, J.M. and C.K. Augspurger. 1999. Geomorphic processes and spatial patterns of primary forest succession on the Bogue Chitto River, USA. Journal of Ecology 87:1052–1063.

Robinson, C.T., K. Tockner, and J.V. Ward. 2002. The fauna of dynamic riverine landscapes. Freshwater Biology 47:661–677.

Rodriguez-Iturbe, I. and A. Rinaldo. 1997. Fractal River Basins: Chance and Self-Organisation. Cambridge University Press, Cambridge, U.K.

Rogers, K.H. 2003. Adopting a heterogeneity paradigm; implications for management of protected savannas. Pages 41–58 *in*: J.T. Du Toit, K.H. Rogers, and H.C. Biggs (eds.), The Kruger Experience: Ecology and Management of Savanna Heterogeneity. Island Press, Washington.

Rogers, K.H. 2006. The real river management challenge: Integrating scientists, stakeholders and service agencies. River Research and Applications 22:269–280.

Rosgen, D.L. 1994. A classification of natural rivers. Catena 22:169–199.

Santoul, F., J. Cayrou, S. Mastrorillo, and R. Céréghino. 2005. Spatial patterns of the biological traits of freshwater fish communities in south-west France. Journal of Fish Biology 66:301–314.

Schade, J.D., S.G. Fisher, N.B. Grimm, and J.A. Seddon. 2001. The influence of a riparian shrub on nitrogen cycling in a Sonoran Desert stream. Ecology 82:3363–3376.

Schaller, J.L., T.V. Royer, M.B. David, and J.L. Tank. 2004. Denitrification associated with plants and sediments in an agricultural stream. Journal of the North American Benthological Society 23:667–676.

Schiemer, F., H. Keckeis, W. Reckendorfer, and G. Winkler. 2001a. The "inshore retention concept" and its significance for large rivers. Archiv für Hydrobiologie, Supplement 135, Large rivers 12:509–516.

Schiemer, F., H. Keckeis, H. Winkler, and L. Flore. 2001b. Large rivers: The relevance of ecotonal structure and hydrological properties for the fish fauna. Archiv für Hydrobiologie, Supplement 135, Large Rivers 12:487–508.

Schilling, E.B. and B.G. Lockaby. 2006. Relationships between productivity and nutrient circulation within two contrasting Southeastern U.S. floodplain forests. Wetlands 26:181–192.

Schlosser, I.J. 1987. A conceptual framework for fish communties in small warmwater streams. Pages 17–24 *in*: W.J. Matthews and D.C. Heins (eds.), Communty and Evolutionary Eecology of North American Stream Fishes. University of Oklahoma Press, Norman, Oklahoma, USA.

Schmidt, J.C., R.A. Parnell, P.E. Grams, J.E. Hazel, M.A. Kaplinski, L.E. Stevens, and T.L. Hoffnagle. 2001. The 1996 controlled flood in Grand Canyon: Flow, sediment transport, and geomorphic change. Ecological Applications 11:657–671.

Schoener, T.W. 1989. Food webs from the small to the large. Ecology 70:1559–1589.

Schofield, N.J., K.J. Collier, J. Quinn, F. Sheldon, and M.C. Thoms. 2000. Australia and New Zealand. Pages 311–334 *in*: P.J. Boon, B.R. Davies, and G.E. Petts (eds.), Global Perspectives on River Conservation: Science, Policy and Practice. Wiley, Chichester.

Schumm, S.A. 1977a. The Fluvial System. Wiley, New York.

Schumm, S.A. 1977b. Patterns of alluvial rivers. Annual Review of Earth and Planetary Sciences 13:5–27.

Schumm, S.A. 1988. Variability of the fluvial system in space and time. Pages 225–250 *in*: T. Rosswall, R.G. Woodmansee, and P.G. Risser (eds.), Scales and global change: spatial and temporal variability in biospheric and geospheric processes. John Wiley and Sons, Chichester.

Schumm, S.A. 1991. To Interpret the Earth, Ten Ways to be Wrong. Cambridge University Press, Cambridge.

Schumm, S.A. and R.W. Lichty. 1965. Time, space, and causality in geomorphology. American Journal of Science 263:110–119.

Scott, M.L., G.T. Auble, and J.M. Friedman. 1997. Flood dependency of cottonwood establishment along the Missouri River, Montana, USA. Ecological Applications 7:677–690.

Sedell, J.R., J.E. Richey, and F.J. Swanson. 1989. The river continuum concept: A basis for the expected behavior of very large rivers? Pages 49–55 *in*: D.P. Dodge (ed.), Proceedings of the International Large River Symposium. Canadian Special Publications in Fishers Aquatic Sciences 106.

Sellers, T. and P.A. Bukaveckas. 2003. Phytoplankton production in a large, regulated river: A modeling and mass balance assessment. Limnology and Oceanography 48:1476–1487.

Shakesby, R.A. and S.H. Doerr. 2006. Wildfire as a hydrological and geomorphological agent. Earth-Science Reviews 74:269–307.

Sheldon, F. and M.C. Thoms. 2004. River corridors. Pages 280–297 *in*: R. Breckwodt, R. Boden, and J. Andrews (eds.), The Darling. The Murray Darling Basin Commission.

Sheldon, F. and M.C. Thoms. 2006a. Geomorphic in-channel complexity: The key to organic matter retention in large dryland rivers? Geomorphology 77:275–285.

Sheldon, F. and M.C. Thoms. 2006b. Relationships between flow variability and macroinvertebrate assemblage composition: Data from Australian dryland rivers. River Research and Applications 22:219–238.

Sheldon, F. and K.F. Walker. 1997. Changes in biofilm induced by flow regulation could explain extinctions of aquatic snails in the lower River Murray, Australia. Hydrobiologia 347:97–108.

Sheldon, F., A.J. Boulton, and J.T. Puckridge. 2002. Conservation value of variable connectivity: Aquatic invertebrate assemblages of channel and floodplain habitats of a central Australian arid-zone river, Cooper Creek. Biological Conservation 103:13–31.

Shields, F.D., Jr. and R.H. Smith. 1992. Effects of large woody debris removal on physical characteristics of a sand-bed river. Aquatic Conservation: Marine and Freshwater Ecosystems 2:145–163.

Slipke, J.W., S.M. Simmons, and M.J. Maceina. 2005. Importance of the connectivity of backwater areas for fish production in Demopolis Reservoir, Alabama. Journal of Freshwater Ecology 20:479–485.

Smith, M., H. Cawell, and P. Mettler-Cherry. 2005. Stochastic flood and precipitation regimes and the population dynamics of a threatened floodplain plant. Ecological Applications 15:1036–1052.

Snelder, T.H. and B.J.F. Biggs. 2002. Multi-scale river environment classification for water resource management. American Water Resources Association 38:1225–1239.

Sobczak, W.V. 2005. Lindeman's trophic dynamic aspect of ecology: "Will you still need me when I'm 64?" Limnology and Oceanography Bulletin 14:53–57.

Sommer, T.R., R.D. Baxter, and F. Feyrer. 2007. Splittail "delisting": a review of recent population trends and restoration activities. American Fisheries Society Symposium 53:25–38.

Southwood, T.R.E. 1977. Habitat, the templet for ecological strategies? Journal of Animal Ecology 46:337–365.

Sparks, R.E. 1995. Need for ecosystem management of large rivers and their floodplains. Bioscience 45:168–182.

Sparks, R.E., P.B. Bayley, S.L. Kohler, and L.L. Osborne. 1990. Disturbance and recovery of large floodplain rivers. Environmental Management 14:699–709.

Sparks, R.E., J.C. Nelson, and Y. Yin. 1998. Naturalization of the flood regime in regulated rivers. Bioscience 48:706–720.

Stanford, J.A. 2006. Chapter 1: Landscapes and rivescapes. Pages 3–21 *in*: F.R Hauer and G.A. Lamberti (eds.), Methods in Stream Ecology. Elsevier, New York.

Stanford, J.A. and J.V. Ward. 1993. An ecosystem perspective of alluvial rivers: Connectivity and the hyporheic corridor. Journal of the North American Benthological Society 12:48–60.

Stanford, J.A., M.S. Lorang, and F.R. Hauer. 2005. The shifting habitat mosaic of river ecosystems. Verhandlungen der Internationalen Vereinigung für Theoretische und Angewandte Limnologie 29:123–126.

Stanford, J.A., J.V. Ward, W.J. Liss, C.A. Frissell, R.N. Williams, J.A. Lichatowich, and C.C. Coutant. 1995. A general protocol for restoration of regulated rivers. Regulated Rivers: Research and Management 12:391–413.

Stanley, E.H. and A.D. Boulton. 2000. River size as a factor in conservation. Pages 403–414 *in*: P.J. Boon, B.R. Davies, and G.E. Petts (eds.), Global perspectives on river conservation. Wiley, Chichester.

Starry, O.S., H.M. Valett, and M.E. Schreiber. 2005. Nitrification rates in a headwater stream: Influences of seasonal variation in C and N supply. Journal of the North American Benthological Society 24:753–768.

Starzecka, A. and T. Bednarz. 2001. Metabolism of bottom sediments in littoral of submountain dam reservoir and the river feeding it. International Journal of Ecohydrology and Hydrobiology 1:441–447.

Strayer, D.L., H.M. Malcolm, R.E. Bell, S.M. Carbotte, and F.O. Nitsche. 2000. Using geophtsical information to define benthic habitats in a large river. Freshwater Biology 51:25–48.

Statzner, B. 1981. Shannon-Weaver diversity of the macrobenthos in the Schierenseebrooks (North Germany) and problems of its use for the interpretation of the community structure. Verhein Internationale Vereinigung Limnologie 21:782–786.

Statzner, B. and B. Higler. 1985. Questions and comments on the river continuum concept. Canadian Journal of Fisheries and Aquatic Sciences 42:1038–1044.

Statzner, B. and B. Higler. 1986. Stream hydraulics as a major determinant of benthic invertebrate zonation patterns. Freshwater Biology 16: 127–139.

Statzner, B., P. Sagnes, J.Y. Chanpagne, and S. Viboud. 2003. Contribution of benthic fish to the patch dynamics of gravel and sand transport in streams. Water Resources Research 39(11):1309, doi:10.1029/2003WR002270.

Steffan, A.W. 1971. Chironomid (Diptera) biocoenoses in Scandinavian glacier brooks. Canadian Entomologist 103:477–486.

Stevens, L.E., J.P., Shannon, and D.W. Blinn. 1997. Colorado River benthic ecology in Grand Canyon, Arizona, USA: Dam, tributary, and geomorphological influences. Regulated Rivers: Research and Management 13:129–149.

Stewart, J.G., C.S. Schieble, R.C. Cashner, and V.A. Barko. 2005. Long-term trends in the Bogue Chitto River fish assemblage: A 27 year perspective. Southeastern Naturalist 4:261–272.

Stoeckel, J.A., D.W. Schneider, L.A. Soeken, K.D. Blodgett, and R.E. Sparks. 1997. Larval dynamics of a riverine metapopulation: Implications for zebra mussel recruitment, dispersal, and control in a large-river system. Journal of the North American Benthological Society 16:586–601.

Strahler, A.N. 1957. Quantitative analysis of watershed geomorphology. American Geophysical Union Transactions 8:913–920.

Strauss, E.A., W.B. Richardson, J.C. Cavanaugh, L.A. Bartsch, R.M. Kreiling, and A.J. Standorf. 2006. Variability and regulation of denitrification in an Upper Mississippi River backwater. Journal of the North American Benthological Society 25:596–606.

Tayfur, G. and V. Guldal. 2006. Artificial neural networks for estimating daily total suspended sediment in natural streams. Nordic Hydrology 37:69–79.

Tennant, D.L. 1976. Instream flow regimes for fish, wildlife, recreation and related environmental resources. Fisheries 1:6–10.

Tharme, R.E. 2003. A global perspective on environment flow assessment: Emerging trends in the development and application environmental flow methodologies for rivers. River Research and Applications 19:397–441.

Thompson, R. and C. Townsend. 2005a. Food-web topology varies with spatial scale in a patchy environment. Ecology 86:1916–1925.

Thompson, R.M. and C.R. Townsend. 2005b. Energy availability, spatial heterogeneity and ecosystem size predict food-web structure in streams. Oikos 108:137–148.

Thompson, R.M and C.R Townsend. 2006. A truce with neutral theory: Local deterministic factors, species traits and dispersal limitation together determine patterns of diversity in stream invertebrates. Journal of Animal Ecology 75:476–484.

Thomaz, S.M., L.M. Bini, and R.L. Bozelli. 2007. Floods increase similarity among aqautic habitats in river-floodplain systems. Hydrobiologia 579:1–13.

Thoms, M.C. 2003. Floodplain-river ecosystems: Lateral connections and the implications of human interference. Geomorphology 56:335–349.

Thoms, M.C. 2006. Variability in riverine ecosystems. River Research and Applications 22:115–122.

Thoms, M.C., S.M. Hill, M.J. Spry, X.J. Chen, T.J. Mount, and F. Sheldon. 2004. The geomorphology of the Darling River. Pages 68–105 in: R. Breckwodt, R. Boden, and J. Andrews (eds.), The Darling. The Murray Darling Basin Commission.

Thoms, M.C. and M. Parsons. 2002. Ecogeomorphology: An interdisciplinary approach to river science. International Association of Hydrological Sciences 27:113–119.

Thoms, M.C. and M. Parsons. 2003. Identifying spatial and temporal patterns in the hydrological character of the Condamine-Balonne River, Australia, using multivariate statistics. River Research and Applications 19:443–457.

Thoms, M.C., S. Rayburg, and M. Neave. 2007. The physical diversity and assessment of a large river system: The Murray-Darling Basin, Australia. Pages 587–608 in: A. Gupta (ed.) Large rivers. Wiley, Chichester.

Thoms, M.C. and F. Sheldon. 1997. River channel complexity and ecosystem processes: The Barwon-Darling River (Australia). Pages 193–205 in: N. Klomp and I. Lunt (eds.), Frontiers in ecology: Building the links. Elsevier.

Thoms, M.C. and F. Sheldon. 2000. Water resource development and hydrological change in a large dryland river system: The Barwon-Darling River, Australia. Journal of Hydrology 228:10–21.

Thoms, M.C. and F. Sheldon. 2002. An ecosystem approach for determining environmental water allocations in Australian dryland river systems: The role of geomorphology. Geomorphology 47:153–168.

Thoms, M.C., M.R. Southwell, and H.M. McGinness. 2005. Water resource development and the fragmentation of floodplain river ecosystems. Geomorphology 71:126–138.

Thoms, M.C, P. Suter, J. Roberts, J. Koehn, G. Jones, T. Hillman, and A. Close. 1998. River Murray scientific panel on environmental flows: River Murray from Dartmouth to Wellington and the Lower Darling. Murray Darling Basin Commission, 450p.

Thoms, M.C. and K.F. Walker. 1993. A case history of the environmental effects of flow regulation on a semi-arid lowland river: The River Murray, South Australia. Regulated Rivers: Research and Management, 8:103–119.

Thorp, J.H. 1986. Two distinct roles for predators in freshwater assemblages. Oikos 47:75–82.

Thorp, J.H. 1992. Linkage between islands and benthos in the Ohio River, with implications for riverine management. Canadian Journal of Fisheries and Aquatic Sciences 49:1873–1882.

Thorp, J.H. and A.F. Casper. 2002. Potential effects on zooplankton from species shifts in mussel planktivory: A field experiment in the St. Lawrence River. Freshwater Biology 47:107–119.

Thorp, J.H. and A.F. Casper. 2003. Importance of biotic interactions in large rivers: An experiment with planktivorous fish, dreissenid mussels, and zooplankton in the St. Lawrence. River Research and Applications 19:265–279.

Thorp, J.H. and M.L. Cothran. 1984. Regulation of freshwater community structure at multiple intensities of dragonfly predation. Ecology 65:1546–1555.

Thorp, J. H. and A. P. Covich (eds.). 2001. Ecology and Classification of North American Freshwater Invertebrates. Second Edition, Academic Press, San Diego. 1038p.

Thorp, J.H. and M.D. Delong. 1994. The riverine productivity model: An heuristic view of carbon sources and organic processing in large river ecosystems. Oikos 70:305–308.

Thorp, J.H. and M.D. Delong. 2002. Dominance of autochthonous autotrophic carbon in food webs of heterotrophic rivers? Oikos 96:543–550.

Thorp, J.H., M.D. Delong, and A.F. Casper. 1998. *In situ* experiments on predatory regulation of a bivalve mollusc (*Dreissena polymorpha*) in the Mississippi and Ohio Rivers. Freshwater Biology 39:649–661.

Thorp, J.H. and S. Mantovani. 2005. Zooplankton in turbid and hydrologically dynamic, prairie rivers. Freshwater Biology 50:1474–1491.

Thorp, J.H., M.C. Thoms, and M.D. Delong. 2006. The riverine ecosystem synthesis: Biocomplexity in river networks across space and time. River Research and Applications 22:123–147.

Tibbetts, T.M. and M.C. Molles. 2005. C:N:P stoichiometry of dominant riparian trees and arthropods along the Middle Rio Grande. Freshwater Biology 50:1882–1894.

Tiemann, J.S., D.P. Gillette, M.L. Wildhaber, and D.R. Edds. 2005. Effects of lowhead dams on the ephemeropterans, plecopterans, and trichopterans group in a North American river. Journal of Freshwater Ecology 20:519–525.

Tockner, K., S.E. Bunn, G. Quinn, R. Naiman, J.A. Stanford, and C. Gordon. 2007. Floodplains: Critically threatened ecosystems, Pages xxx-xxx *in*: Polunin, N.C. (eds), The State of the World's Ecosystems. Cambridge University Press.

Tockner, K., U. Karaus, A. Paetzold, C. Claret, and J. Zettel. 2006. Ecology of braided rivers. Pages 339–360 *in*: G. Sambrook Smith, J. Best, C. Bristow, and G.E. Petts (eds.), Braided Rivers. IAS Special Publication, Blackwell, Oxford, UK.

Tockner, K., F. Malard, and J.V. Ward. 2000. An extension of the flood pulse concept. Hydrological Processes 14:2861–2883.

Tockner, K., D, Pennetzdorfer, N. Reiner, F. Schiemer, and J.V. Ward. 1999a. Hydrological connectivity, and the exchange of organic matter and nutrients in a dynamic river-floodplain system (Danube, Austria). Freshwater Biology 41:521–535.

Tockner, K., F. Schiemer, C. Baumgartner, G. Kum, E. Wiegand, I. Zweimueller, and J.V. Ward. 1999b. The Danube Restoration Project: Species diversity patterns across connectivity gradients in the floodplain system. Regulated Rivers: Research and Management 15:245–258.

Tockner, K., F. Schiemer, and J.V. Ward. 1998. Conservation by restoration: The management concept for a river-floodplain system on the Danube River in Austria. Aquatic Conservation: Marine and Freshwater Ecosystems 8:71–86.

Tokeshi, M. 1993. Community ecology and patchy freshwater habitats. Pages 63–91 *in*: P.S. Giller, A.G. Hildrew, and D.G. Raffaelli (eds.), Aquatic Ecology: Scale, Pattern and Process. Blackwell Scientific Publications, London.

Townsend, C.R. 1989. The patch dynamics concept of stream community ecology. Journal of the North American Benthological Society 8:36–50; 14:2861–2883.

Townsend, C.R. 1996. Concepts in river ecology: Pattern and process in the catchment hierarchy. Archiv für Hydrobiologie, Supplement 113, Large Rivers 10:3–21.

Townsend, C.R. and A.G. Hildrew. 1994. Species traits in relation to a habitat templet for river systems. Freshwater Biology 31:265–275.

Townsend, C.R., S. Dolédec, and M.R. Scarsbrook. 1997. Species traits in relation to temporal and spatial heterogeneity in streams: A test of habitat templet theory. Freshwater Biology 37:367–387.

Tucker, J.K., C.H. Theiling, and J.B. Camerer. 1996. Utilization of backwater habitats by unionid mussels (Bivalvia: Unionidae) on the lower Illinois River and in Pool 26 of the Upper Mississippi River. Transactions of the Illinois State Academy of Science 89:113–122.

Turak, E. 2007. An Ecological Typology of the Rivers of New South Wales, Australia. Unpublished PhD thesis, University of Technology, Sydney, Australia.

Turner, M.G., W.H. Romme, R.H. Gardner, R.V. O'Neill, and T.K. Kratz. 1993. A revised concept of landscape equilibrium – disturbance and stability on scaled landscapes. Landscape Ecology 8:213–227.

Uehlinger, U. 2000. Resistance and resilience of ecosystem metabolism in a flood-prone river system. Freshwater Biology 45:319–332.

Uehlinger, U. 2006. Annual cycle and inter-annual variability of gross primary production and ecosystem respiration in a floodprone river during a 15-year period. Freshwater Biology 51:938–950.

Valdez, R.A., T.L. Hoffnagle, C.C. McIvor, T. McKinney, and W.C. Leibfried. 2001. Effects of a test flood on fishes of the Colorado River in Grand Canyon, Arizona. Ecological Applications 11:686–700.

Valett, H.M., C.L. Crenshaw, and P.F. Wagner. 2002. Stream nutrient uptake, forest succession, and biogeochemical theory. Ecology 83:2888–2901.

Vander Zanden, M.J., S. Chandra, B.C. Allen, J.E. Reuter, and C.R. Goldman. 2003. Historical food web structure and restoration of native aquatic communities in the Lake Tahoe (California-Nevada) Basin. Ecosystems 6:274–288.

Vander Zanden, M.J. and J.B. Rasmussen. 1996. A trophic position model of pelagic food webs: Impact on contaminant bioaccumulation in lake trout. Ecological Monographs 66:451–477.

Vander Zanden, M.J. and J.B. Rasmussen. 1999. Primary consumer δ^{13}C and δ^{15}N and the trophic position of aquatic consumers. Ecology 80:1395–1404.

Vander Zanden, M.J., B.J. Shuter, N. Lester, and J.B. Rasmussen. 1999. Patterns of food chain length in lakes: A stable isotope study. American Naturalist 154:406–416.

Van der Nat, D., K. Tockner, P.J. Edwards, and J.V. Ward. 2003. Large woods dynamics of complex Alpine river floodplains. Journal of the North American Benthological Society 22:35–50.

Van Geest, G.J., H. Coops, R.M.M. Roijackers, A.D. Buijse, and M. Scheffer. 2005. Succession of aquatic vegetation driven by reduced water-level fluctuations in floodplain lakes. Journal of Applied Ecology 42:251–260.

Van Niekerk, A.W., G.L. Heritage, and B.P. Moon. 1995. River classification for management: The geomorphology of the Sabie River in the Eastern Transvaal. South African Geographical Journal 77:68–76.

Vannote, R.L., G.W. Minshall, K.W., Cummins, J.R., Sedell, and C.E. Cushing. 1980. The river continuum concept. Canadian Journal of Fisheries and Aquatic Sciences 37:130–137.

Veitzer, S.A., B.M. Troutman, and V.K. Gupta. 2003. Power-law tail probabilities of drainage areas in river basins. Physical Review E 68:016123.

Vranovsky, M. 1995. The effects of current velocity upon the biomass of zooplankton in the River Danube side arms. Biologia 50:461–464.

Walker, K.F., F. Sheldon, and J.T. Puckridge. 1995. A perspective on dryland river ecosystems. Regulated Rivers: Research and Management 11:85–104.

Walters, C.J. 1986. Adaptive management of renewable resources. Macmillan Publishing Company, New York. 374 p.

Walters, C.J. and J., Korman. 1999. Cross-scale modelling of riparian ecosystem responses to hydrological management. Ecosystems 2:411–421.

Walters, D.M., K.M. Fritz, and D.L. Phillips. 2007. Reach-scale geomorphology affects organic matter and consumer δ^{13}C in a forested Piedmont stream. Freshwater Biology 52:1105–1119.

Walters, D.M., D.S. Leigh, M.C. Freeman, and B.J. Freeman. 2003. Geomorphology and fish assemblages in a Piedmont river basin, USA. Freshwater Biology 48:1950–1970.

Wantzen, K.M. 2006. Physical pollution: Effects of gully erosion on benthic invertebrates of a tropical clear-water stream. Aquatic Conservation 16:733–749.

Ward, J.V. 1989. The four-dimensional nature of lotic ecosystems. Journal of the North American Benthological Society 8: 2–8; 37:130–137.

Ward, J.V. and J.A. Stanford. 1983a. The intermediate disturbance hypothesis: An explanation for biotic diversity patterns in lotic ecosystems. Pages 347–356 *in*: T.D. Fontaine and S.M. Bartell (eds.), Dynamics of lotic ecosystems, Ann Arbor Science Publishers, Ann Arbor, MI, USA.

Ward, J.V. and J.A. Stanford. 1983b. The serial discontinuity concept of lotic ecosystems. Pages 43–68 *in*: T.D. Fontaine, III and S.M. Bartell (eds.), Dynamics of Lotic Ecosystems. Ann Arbor Science, Ann Arbor, Michigan, USA.

Ward, J.V. and K. Tockner. 2001. Biodiversity: Towards a unifying theme for river ecology. Freshwater Biology 46:807–819.

Ward, J.V., K. Tockner, D.B. Arscott, and C. Claret. 2002. Riverine landscape diversity. Freshwater Biology 47:57–539.

Ward, J.V., K. Tockner, and F. Schiemer. 1999. Biodiversity of floodplain river ecosystems: Ecotones and connectivity. Regulated Rivers: Research and Management. 15:125–139.

Ward, J. V. and U. Uehlinger (eds.) 2003. Ecology of a Glacial Floodplain. Kluwer Academic Publisher, Dordrecht, NL.

Ward, J.V. and N.J. Voelz. 1994. Groundwater fauna of the South Platte River system, Colorado. Pages 391–423 *in*: J. Gibert, D.L. Danielpol, and J.A. Stanford (eds.), Groundwater Ecology. Academic Press, San Diego.

Warner, R.F. 1987. Spatial adjustments to temporal variations in flood regime in some Australian Rivers. Pages 14–40 *in*: K.S. Richards (ed.), Rivers: Environment and Process. Blackwell, Oxford.

Webb, A.A. and W.D. Erskine. 2003. A practical scientific approach to riparian vegetation rehabilitation in Australia. Journal of Environmental Management 68:329–341.

Webster, J.R. 2007. Spiraling down the river continuum: Stream ecology and the U-shaped curve. Journal of the North American Benthological Society 26:375–389.

Webster, J.R. and E.F. Benfield. 1986. Vascular plant breakdown in freshwater ecosystems. Annual Review of Ecology and Systematics 17:567–594.

Webster, J.R. and B.C. Patten. 1979. Effects of watershed perturbation on stream potassium and calcium dynamics. Ecological Monographs 49:51–72.

Weigel, B.M., J. Lyons, and P.W. Rasmussen. 2006. Fish assemblages and biotic integrity of a highly modified floodplain river, the Upper Mississippi, and a large, relatively unimpacted tributary, the lower Wisconsin. River Research and Applications 22:923–936.

Werner, B.T., 1999. Complexity in Natural Landform Patterns. Science 284:102–104.

Wetzel, R.G. 2001. Limnology: Lake and River Ecosystems. Third edition., Academic Press, San Diego, CA.

White, D., K. Johnston, and M. Miller. 2005. Ohio River basin. Pages 375–424 *in*: A.C. Benke and C.E. Cushing (eds.), River of North America. Elsevier, London, UK.

White, P.S. and J. Harrod. 1997. Disturbance and diversity in a landscape context. Pages 955–970 *in*: J.A. Bissonette (ed.), Wildlife and landscape ecology: Effects of pattern and scale. Springer, New York.

White, P.S. and S.T.A. Pickett. 1985. Natural disturbance and patch dynamics: An introduction. Chapter 1, Pages 3–13 *in*: S.T.A. Pickett and P.S. White (eds.), The Ecology of Natural Disturbance and Patch Dynamics. Academic Press, New York. 472 p.

Whitley, J.R. and R.S. Campbell. 1974. Some aspects of water quality and biology of the Missouri River. Transactions of the Missouri Academy of Science 8:60–72.

Wiederholm, T. and R.K. Johnson. 1997. Monitoring and assessment of lakes and watercourses in Sweden. Pages 317–329 *in*: J.J. Ottens, F.A.M. Claessen, P.G. Stoks, J.G. Timmerman, and R.C. Ward (eds.), Monitoring Tailor-made II, Information strategies in Water, Nunspeet, The Netherlands.

Wiens, J.A. 1989. Spatial scaling in ecology. Functional Ecology 3:385–397.

Wiens, J.A. 2002. Riverine landscapes: Taking landscape ecology into the water. Freshwater Biology 47:501–515.

Wilczek, S., H. Fischer, and M.T. Pusch. 2005. Regulation and seasonal dynamics of extracellular activities of a large lowland river. Microbial Ecology 50:253–267.

Winemiller, K.O 2005. Floodplain river food webs: Generalizations and implications for fisheries management. Pages 285–312 *in*: R.L. Welcomme and T. Petr (eds.), Proceedings of the Second International Symposium on the Management of Large Rivers for Fisheries, Volume 2. Mekong River Commission, Phnom Penh, Cambodia.

Winemiller, K.O., S. Akin, and S.C. Zeug. 2007. Production sources and food web structure of a temperate tidal estuary: integration of dietary and stable isotope data. Marine Ecology Progress Series 343:63–76.

Winemiller, K.O., J.V. Montoya, D.L. Roelke, C.A. Layman, and J.B. Cotner. 2006. Seasonally varying impact of detritivorous fishes on the benthic ecology of a tropical floodplain river. Journal of the North American Benthological Society 25:250–262.

Winston, M.R., C.M. Taylor, and J. Pigg. 1991. Upstream extirpation of four minnow species due to damming of a prairie stream. Transactions of the American Fisheries Society 120:98–105.

Winterbourn, M.J., J.S. Rounick, and B. Cowie. 1981. Are New Zealand stream ecosystems really different? New Zealand Journal of Marine and Freshwater Research 15:321–328.

Wissmar, R.C. and P.A. Bisson. 2003. Strategies for restoring river ecosystems: Sources of variability and uncertainty in natural and managed systems. American Fisheries Society, Bethesda, Maryland.

Wondzell, S.M. and F.J. Swanson. 1996. Seasonal and storm dynamics of the hyporheic zone of a 4th-order mountain stream. I: Hydrologic processes. Journal of the North American Benthological Society 15:3–19.

Wondzell, S.M. and F.J. Swanson. 1999. Floods, channel change, and the hyporheic zone. Water Resources Research 35:555–567.

Woodward, G. and A.G. Hildrew. 2002. Food web structure in riverine landscapes. Freshwater Biology 47:777–798.

Woodward, G., R. Thompson, C.R. Townsend, and A.G. Hildrew. 2005. Pattern and process in food webs: Evidence from running waters. Pages 51–66 *in*: A. Belgrano, U.M. Scharler, J. Dunne, and R.E. Ulanowicz. 2005. Aquatic food webs: An ecosystem approach. Oxford University Press, Oxford, UK.

Wootton, J.T. 1998. Effects of disturbance on species diversity: A multitrophic perspective. American Naturalist 152:803–825.

Wright, J.F. 1995. Development and use of a system for predicting the macroinvertebrate fauna in flowing waters. Australian Journal of Ecology 20:181–197.

Wright, J.F. 2000. An Introduction to RIVPACS. Pages 1–24 *in*: J.F. Wright, D.W. Sutcliffe, and M.T. Furse (eds.), Assessing the Biological Quality of Fresh Waters: RIVPACS and Other Techniques. Freshwater Biological Association, Ambleside, U.K.

Wu, J. 1999. Hierarchy and scaling: Extrapolating information along a scaling ladder. Canadian Journal of Remote Sensing 25:367–380.

Wu, J. and O.L. Loucks. 1995. From balance of nature to hierarchical patch dynamics: A paradigm shift in ecology. Quarterly Review of Biology 70:439–466.

Ziegler, A.D., T.W. Giambelluca, R.A. Sutherland, M.A. Nullet, S. Yarnasarn, J. Pinthong, P. Preechapanya, and S. Jaiaree. 2004. Toward understanding the cumulative impacts of roads in upland agricultural watersheds of northern Thailand. Agriculture, Ecosystems & Environment 104:145–158.

Index